Low-Power Design of Nanometer FPGAs

Architecture and EDA

The Morgan Kaufmann Series in Systems on Silicon

Series Editor: Wayne Wolf, Georgia Institute of Technology

The Designer's Guide to VHDL, Second Edition
Peter J. Ashenden

The System Designer's Guide to VHDL-AMS
Peter J. Ashenden, Gregory D. Peterson, and Darrell A. Teegarden

Modeling Embedded Systems and SoCs
Axel Jantsch

ASIC and FPGA Verification: A Guide to Component Modeling
Richard Munden

Multiprocessor Systems-on-Chips
Edited by Ahmed Amine Jerraya and Wayne Wolf

Functional Verification
Bruce Wile, John Goss, and Wolfgang Roesner

Customizable and Configurable Embedded Processors
Edited by Paolo Ienne and Rainer Leupers

Networks-on-Chips: Technology and Tools
Edited by Giovanni De Micheli and Luca Benini

VLSI Test Principles & Architectures
Edited by Laung-Terng Wang, Cheng-Wen Wu, and Xiaoqing Wen

Designing SoCs with Configured Processors
Steve Leibson

ESL Design and Verification
Grant Martin, Andrew Piziali, and Brian Bailey

Aspect-Oriented Programming with e
David Robinson

Reconfigurable Computing: The Theory and Practice of FPGA-Based Computation
Edited by Scott Hauck and André DeHon

System-on-Chip Test Architectures
Edited by Laung-Terng Wang, Charles Stroud, and Nur Touba

Verification Techniques for System-Level Design
Masahiro Fujita, Indradeep Ghosh, and Mukul Prasad

VHDL-2008: Just the New Stuff
Peter J. Ashenden and Jim Lewis

On-Chip Communication Architectures: System on Chip Interconnect
Sudeep Pasricha and Nikil Dutt

Embedded DSP Processor Design: Application Specific Instruction Set Processors
Dake Liu

Processor Description Languages: Applications and Methodologies
Edited by Prabhat Mishra and Nikil Dutt

Three-Dimensional Integrated Circuit Design
Vasilis F. Pavlidis and Eby G. Friedman

Electronic Design Automation: Synthesis, Verification, and Test
Edited by Laung-Terng Wang, Kwang-Ting (Tim) Cheng, Yao-Wen Chang

Design for Debugging and Validation of Embedded Systems
Abhik Roychoudhury

Low-Power Design of Nanometer FPGAs
Hassan Hassan and Mohab Anis

Low-Power Design of Nanometer FPGAs

Architecture and EDA

Hassan Hassan

Mohab Anis

ELSEVIER

AMSTERDAM • BOSTON • HEIDELBERG • LONDON
NEW YORK • OXFORD • PARIS • SAN DIEGO
SAN FRANCISCO • SINGAPORE • SYDNEY • TOKYO

Morgan Kaufmann Publishers is an imprint of Elsevier

MORGAN KAUFMANN

Morgan Kaufmann Publishers is an imprint of Elsevier.
30 Corporate Drive, Suite 400, Burlington, MA 01803, USA

This book is printed on acid-free paper. ♾

Library of Congress Cataloging-in-Publication Data
Application Submitted

British Library Cataloguing-in-Publication Data
A catalogue record for this book is available from the British Library.

ISBN: 978-0-12-374438-8

For information on all Morgan Kaufmann publications, visit our Web site at *www.mkp.com* or *www.elsevierdirect.com*

Typeset by: diacriTech, India

Printed in the United States of America
10 11 12 13 5 4 3 2 1

To my wife, Allaa, and my supportive parents, Abdel Rahman and Nabila
– **Hassan Hassan**

To my wonderful family, Heba, Selim, and Adham
– **Mohab Anis**

Contents

Author Bios

Hassan Hassan received the B.Sc. degree (with honors) in electronics and communication engineering from Cairo University, Cairo, Egypt, in 2001 and the M.A.Sc. and Ph.D. degrees in electrical engineering from the University of Waterloo, Waterloo, ON, Canada, in 2004 and 2008, respectively.

Dr. Hassan is currently a staff engineer in the timing and power group at Actel Corporation. He has authored/coauthored more than 20 papers in international journals and conferences. His research interests include integrated circuit design and design automation for deep submicron VLSI systems. He is also a member of the program committee for several IEEE conferences.

Mohab Anis received the B.Sc. degree (with honors) in electronics and communication engineering from Cairo University, Cairo, Egypt, in 1997 and the M.A.Sc. and Ph.D. degrees in electrical engineering from the University of Waterloo, Waterloo, ON, Canada, in 1999 and 2003, respectively.

Dr. Anis is currently a tenured Associate Professor at the Department of Electrical and Computer Engineering, University of Waterloo. During 2009, he was with the Electronics Engineering Department at the American University in Cairo. He has authored/coauthored over 80 papers in international journals and conferences and is the author of the book *Multi-Threshold CMOS Digital Circuits—Managing Leakage Power* (Kluwer, 2003). His research interests include integrated circuit design and design automation for VLSI systems in the deep submicrometer regime.

Dr. Anis is an Associate Editor of the *IEEE Transactions on Circuits and Systems—II, Microelectronics Journal, Journal of Circuits, Systems and Computers, ASP Journal of Low*

Power Electronics, and *VLSI Design*. He is also a member of the program committee for several IEEE conferences. He was awarded the 2009 Early Research Award, the 2004 Douglas R. Colton Medal for Research Excellence in recognition of excellence in research leading to new understanding and novel developments in microsystems in Canada and the 2002 International Low-Power Design Contest.

Dr. Anis also holds two business degrees: an M.B.A. with concentration in Innovation and Entrepreneurship from Wilfrid Laurier University, and an M.M.S. from the University of Waterloo. He is the co-founder of Spry Design Automation Inc. and has published a number of papers on technology transfer.

Chapter 1

FPGA Overview: Architecture and CAD

Low-Power Design of Nanometer FPGAs: Architecture and EDA

1.1 INTRODUCTION

Field programmable gate arrays (FPGAs) were first introduced to the very-large-scale integration (VLSI) market in the 1980s [1]. Initially, FPGAs were designed to complement application-specific integrated circuit (ASIC) designs by providing reprogrammability on the expense of power dissipation, chip area, and performance. The main advantages of FPGAs compared to ASIC designs can be summarized as follows:

- FPGAs' time-to-market is minimal when compared to ASIC designs. Once a fully tested design is available, the design can be burned into the FPGA and verified, hence initiating the production phase. As a result, FPGAs eliminate the fabrication wait time.
- FPGAs are excellent candidates for low-volume productions since they eliminate the mask generation cost.
- FPGAs are ideal for prototyping purposes. Hardware testing and verification can be quickly performed on the chip. Moreover, design errors can be easily fixed without incurring any additional hardware costs.
- FPGAs are versatile and reprogrammable, thus allowing them to be used in several designs at no additional costs.

Fueled by the increase in the time-to-market pressures, the rise in ASIC mask and development costs, and increase in the FPGAs' performance and system-level features, more and more traditionally ASIC designers are migrating their designs to programmable logic devices (PLDs). Moreover, PLDs progressed both in terms of resources and performance. The latest FPGAs have come to provide platform solutions that are easily customizable for system connectivity, digital signal processing (DSP), and/or data processing applications. These platform building tools accelerate the time-to-market by automating the system definition and integration phases of the system on programmable chip (SoPC) development.

FPGAs belong to a type of VLSI circuits called PLDs. The first PLD devices developed are the programmable array logic (PAL) devices designed by Monolithic Memories Inc. in 1978. PALs adopt a simple PLD (SPLD) architecture, where functionality is provided by a matrix of AND gates followed by a matrix of OR gates to implement sum-of-products function representation, as shown in Fig. 1.1. It should be noted that PAL devices are limited to only two-level logic functionality. Complex PLDs (CPLDs) succeeded PALs in the PLD market to offer higher-order logic functionality. CPLDs consist of SLPD-like devices that are interconnected using a programmable switch

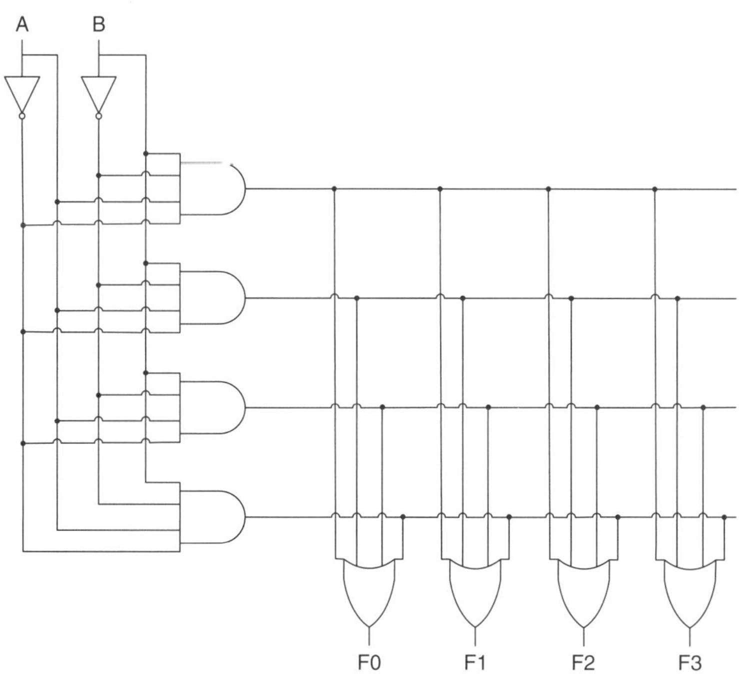

■ **FIGURE 1.1** PAL architecture.

matrix. Despite the increase in the complexity of the functionality of CPLDs, their use remained limited to glue logic in large designs. Finally, FPGAs were introduced to offer more complex functionality by employing a look-up table (LUT) approach to implement logic functions and channel-based routing strategy. The first commercial PLD that adopts the FPGA architecture was developed by Xilinx in 1984.

Recently, FPGA vendors provided a comprehensive alternative to FPGAs for large-volume demands called *structured ASICs* [2, 3]. Structured ASICs offer a complete solution from prototype to high-volume production and maintain the powerful features and high-performance architecture of their equivalent FPGAs with the programmability removed. Structured ASIC solutions not only provide performance improvement, but also result in significant high-volume cost reduction than FPGAs.

FPGAs consist of programmable logic resources embedded in a sea of programmable interconnects. The programmable logic resources can be configured to implement any logic function, while the interconnects provide the flexibility to connect any signal in the design to any logic resource. The programming technology for the logic and interconnect resources can be static random access memory (SRAM), flash memory [4], or antifuse [5, 6]. SRAM-based FPGAs offer in-circuit reconfigurability at the expense of being volatile, while antifuse are write-once devices. Flash-based FPGAs provide an intermediate alternative by providing reconfigurability as well as non-volatility. The most popular programming technology in state-of-the-art FPGAs is SRAM.

Traditionally, FPGAs consist of input/output (IO) pads, logic resources, and routing resources. However, state-of-the-art FPGAs usually include embedded memory, DSP blocks, phase-locked loops (PLLs), embedded processors, analog functionality (e.g., analog-to-digital converters),

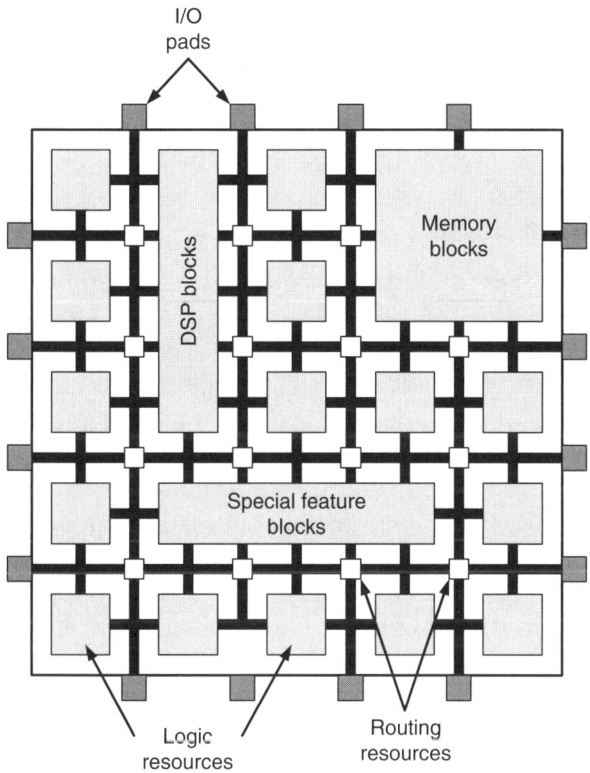

■ **FIGURE 1.2** Modern FPGA fabric

and other special feature blocks, as shown in Fig. 1.2. These features allowed FPGAs to be an attractive alternative for some SoPC designs. The next sections shed light on some of the available commercial FPGA architectures and FPGA CAD flow.

1.2 FPGA LOGIC RESOURCES ARCHITECTURE

The logic blocks in FPGAs are responsible for implementing the functionality needed by each application. Increasing the functional capability of the logic blocks increases the number of logic functions that can be packed into it.

Moreover, increasing the size of logic blocks, i.e., increasing the number of inputs to each logic block, increases the number of logic functions performed by each logic block as well as improving the area/delay performance of the logic block [7]. However, this comes at the expense of wasted resources because not all of the blocks will have all of their inputs fully utilized.

Most commercial FPGAs employ LUTs to implement the logic blocks. A k-input LUT consists of 2^k configuration bits in which the required truth table is programmed during the configuration stage. The almost standard number of inputs for LUTs is four, which was proven optimum for area and delay objectives [8]. However, this number can vary depending on the targeted application by the vendor. Moreover, modern FPGAs utilize a hierarchical architecture, where every group of basic logic blocks are grouped together into a bigger logic structure, logic cluster. The remaining of this section describes the programmable logic resources in three of the most popular commercial FPGAs.

1.2.1 **Altera Stratix IV Logic Resources**

The logic blocks in Altera's Stratix IV are called *adaptive logic modules* (ALMs). An 8-input ALM contains a variety of LUT-based resources that can be divided between two adaptive LUTs [9]. Being adaptive, ALMs can perform the conventional 4-input LUT operations as well as implement any function of up to 6-input and some 7-input functions. Besides the adaptive LUTs, ALMs contain two programmable registers, two dedicated full adders, a carry chain, a shared arithmetic chain, and a register chain. Using these components, ALMs can efficiently perform arithmetic and shift operations. A detailed view of an ALM is shown in Fig. 1.3. Every 8 ALMs are grouped together to form a logic array block (LAB).

To fully benefit from the adaptivity of the ALMs, each ALM can operate in four different modes: normal mode

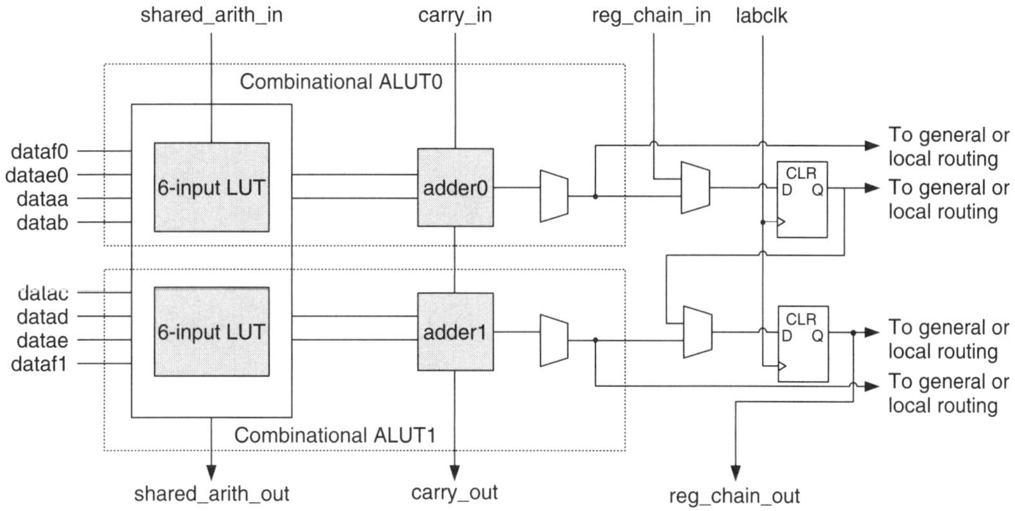

■ **FIGURE 1.3** Altera's Stratix IV ALM architecture [9].

(for general logic applications and combinational functions), extended LUT mode (for implementing some 7-input functions), arithmetic mode (for implementing adders, counters, accumulators, wide parity functions, and comparators), and shared arithmetic mode (for 3-input addition).

1.2.2 **Xilinx Virtex-5 Logic Resources**

The slice is the basic logic resource in Xilinx Virtex-5 FPGAs. Slices consist of four LUTs, wide function multiplexers (MUXs), and carry logic [10]. Figure 1.4 shows the architecture of a typical Virtex-5 slice. The slices employ four 6-LUTs that are capable of performing any 6-input logic function. Functions with up to 8 inputs can be implemented using MUXs to combine the output of two LUTs. Every two interconnected slices are grouped together in a configurable logic block (CLB) [10]. The slice is capable of performing logic, arithmetic, and sequential functionalities.

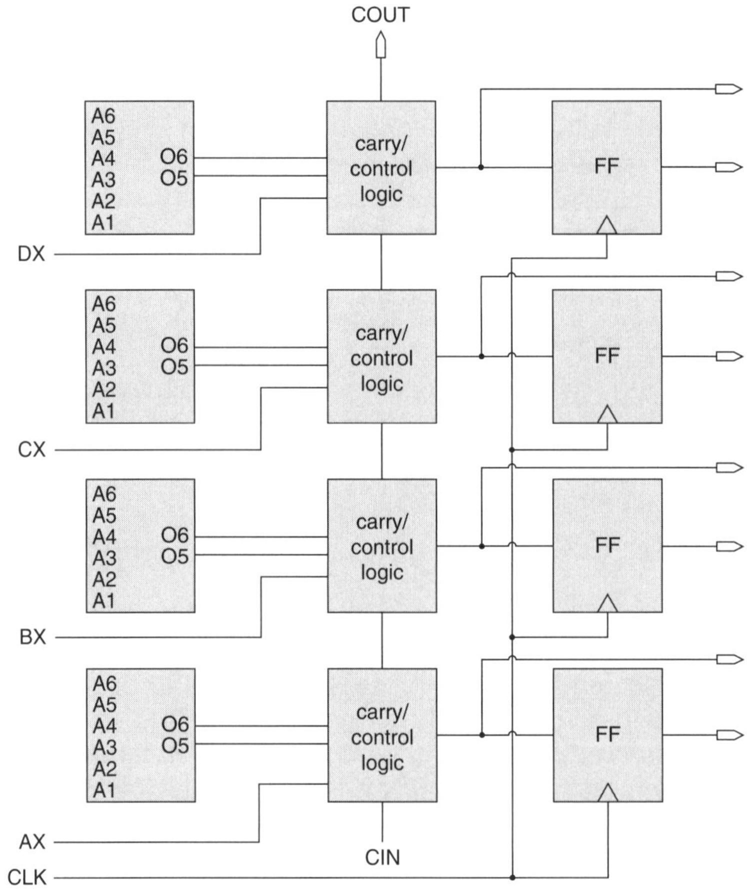

■ **FIGURE 1.4** Xilinx's Vertex-5 slice architecture [10].

1.2.3 Actel ProASIC3/IGLOO Logic Resources

Actel ProASIC3/IGLOO FPGAs employ a flash-based architecture, instead of the conventional SRAM-based FPGAs used by both Altera and Xilinx, to store the configuration bits. The flash architecture provides the FPGAs with both reconfigurability and nonvolatility. The ProASIC3/IGLOO FPGAs employ the VersaTile 3-input logic block that can implement any 3-input logic function as well as sequential functionality, as shown in Fig. 1.5 [11]. Furthermore,

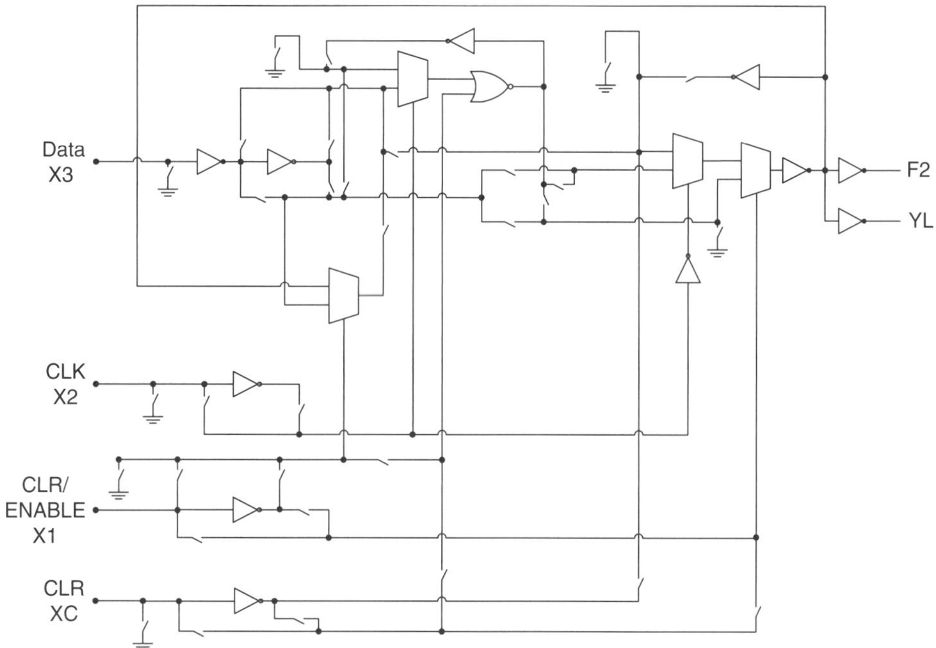

■ **FIGURE 1.5** Actel's ProASIC3/IGLOO VersaTile architecture [11].

the hierarchical architecture is not employed in the ProASIC3/IGLOO FPGAs and the output of each VersaTile can be directly routed to either the fast local lines or the long routing resources. Another interesting characteristic of the VersaTile is that it does not adopt the conventional LUT architecture in FPGAs, as shown in Fig. 1.5. It should be noted that the VersaTile is not capable of performing arithmetic operations since it does not have fast carry chains. Moreover, the ProASIC3/IGLOO FPGAs are not hierarchical in nature and do not have any logic clusters.

1.2.4 **Actel Axcelerator Logic Resources**

Actel Axcelerator FPGA family is an example of nonvolatile permanently programmed FPGA architectures. Antifuse is used to permanently program the FPGA logic cell to

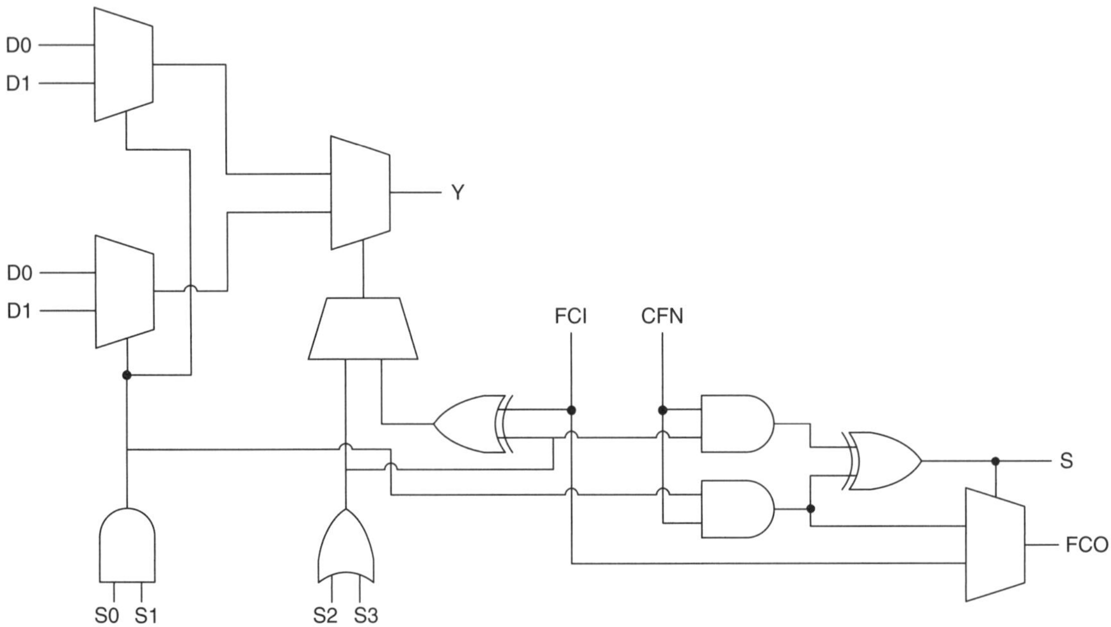

■ **FIGURE 1.6** Actel's Axcelerator C-cell [12].

implement certain functionality. The Axcelerator FPGA is hierarchical with superclusters that contain four C-cells (combinational cells) and two R-cells (sequential cells), as shown in Figs. 1.6 and 1.7. The Axcelerator logic cell can implement all 3-input functions and most 4-input functions, as well as arithmetic operations using the fast carry chain.

1.3 FPGA ROUTING RESOURCES ARCHITECTURE

Routing resources in FPGAs can be divided into two components: segmented local routing and dedicated routing. Segmented local routing is used to provide a connection among the logic blocks. As depicted in Fig. 1.8, the segmented wires are prefabricated in channels to provide programmable connections between switch blocks, connection blocks, and logic blocks. The number of wires in one channel is usually denoted by *W* [13].

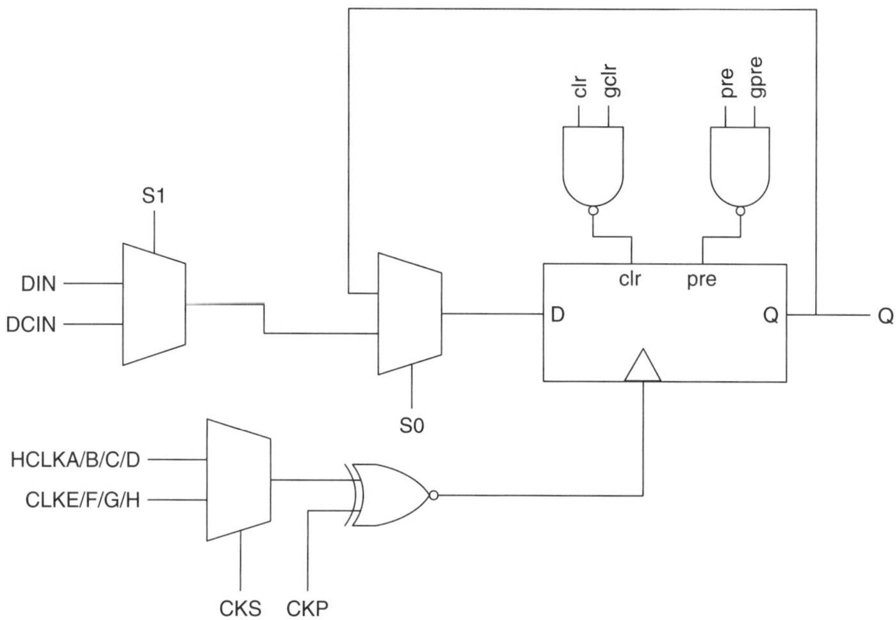

■ **FIGURE 1.7** Actel's Axcelerator R-cell [12].

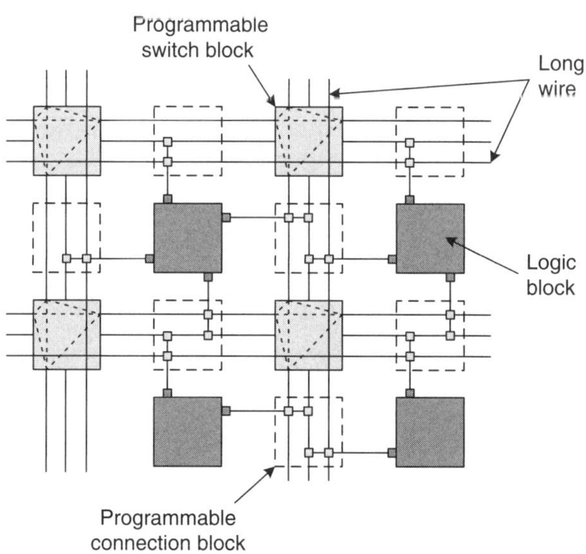

■ **FIGURE 1.8** Routing resources in island-style FPGAs.

The I/O of the logic blocks are dynamically connected to the segmented routing channels on all four sides using connection blocks. The number of wires in each channel to which a logic-block pin can connect is called the *connection block flexibility*, F_C. In addition, the switch blocks provide programmable connectivity between the horizontal and vertical wires. The switch block flexibility F_S is defined as the number of wires to which each incoming wire can connect in a switch block. The segment length of a certain wire segment is defined as the number of logic blocks spanned by the routing wire. Modern FPGAs use a combination of wires of different segment lengths to achieve the optimum performance in terms of routability, delay, or both.

Dedicated routing is used for global signals that fanout to a large number of logic blocks, e.g., clock and reset, thus providing low skew. Moreover, some commercial FPGAs employ PLLs and delay-locked loops (DLLs) for further skew reduction. Modern FPGAs have the flexibility to provide different clock domains inside the FPGA to enable asynchronous designs.

1.4 CAD FOR FPGAs

FPGAs are implemented using a huge number of programmable switches that are used to implement a certain logic function. The CAD tools of FPGAs transform the design, entered either as a schematic or using a hardware description language, to a stream of “1”s and “0”s that program the FPGA during the configuration time. The flowchart in Fig. 1.9 shows the different steps involved in the CAD flow for a typical FPGA design.

1.4.1 Logic Synthesis

In the synthesis phase, the circuit description is converted to a netlist of basic logic gates. This phase is usually divided into two different stages: logic optimization and technology mapping [14–17].

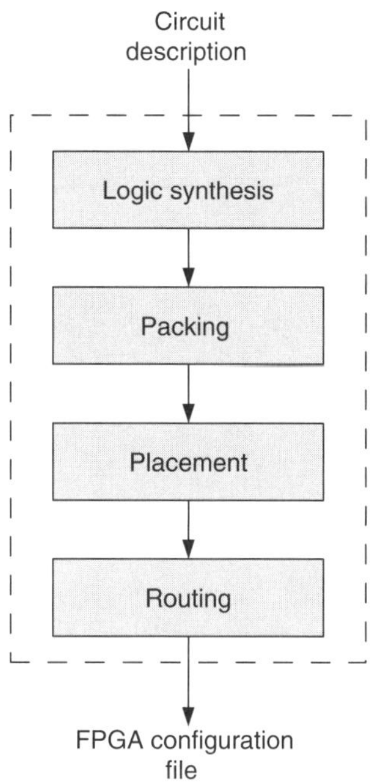

■ **FIGURE 1.9** A typical FPGA CAD flow.

Logic optimization is a technology-independent stage that involves simplifying the logic function of the design without the use of any technology information. Any redundant logic is removed at this stage. The optimized user circuit is then mapped into LUTs and flip-flops in the technology mapping stage, where each k-bounded logic function in the circuit is mapped into a k-LUT. This step resolves to find a set of k-feasible cuts that include all the nodes in the circuit in such a way to minimize the delay, area, and/or power dissipation of the final implementation. The process of technology mapping is often treated as a covering problem.

1.4.2 **Packing**

The packing phase converts the netlist of LUTs and flip-flops into a netlist of logic blocks, as shown in Fig. 1.10.

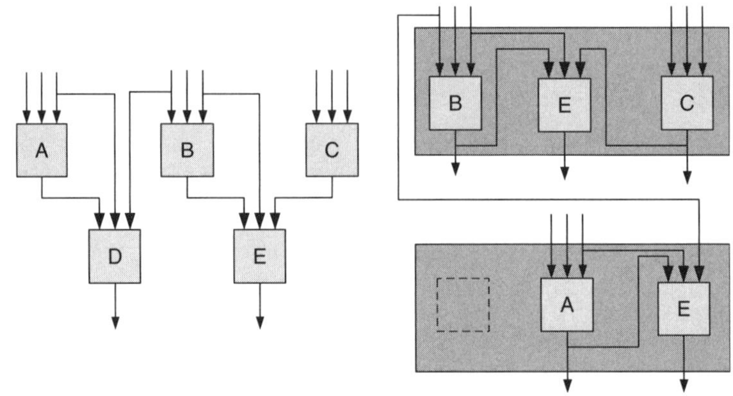

■ **FIGURE 1.10** An example of packing.

The input netlist is converted into clusters of logic blocks that can be mapped into the physical logic blocks of the FPGA. Most packing algorithms minimize the number of resulting logic blocks, the number of connections between them, and/or the delay along the critical path. The packing algorithm has to consider the physical limitations of the actual logic blocks of the FPGA in terms of the maximum number of LUTs in a logic block and the number of distinct input signals and clocks a logic block has.

Packing algorithms can be categorized as either bottom-up [14, 18–20] or top-down [21, 22]. Bottom-up packing algorithms build each cluster individually around a seed LUT until the cluster is full. However, top-down packing algorithms partition the LUTs into clusters by successive circuit subdivision. Bottom-up algorithms are much faster and simpler than top-down algorithms because they only consider local connections. However, this comes at the expense of solution quality.

1.4.3 **Placement**

In the placement phase, the packed logic blocks are distributed among the physical logic blocks in the FPGA fabric. Placement algorithms try to minimize the delay along the critical path and enhance the resulting circuit routability. Available placement algorithms can be classified into three categories: min-cut algorithms [23, 24], analytic algorithms [25, 26], and algorithms based on simulated annealing (SA) [27–29]. Most of the commercial placement tools for FPGAs employ SA-based algorithms due to their flexibility to adapt to a wide variety of optimization goals.

SA placement tools depend on the SA algorithm, which is derived from the annealing process used to cool molten metals [30]. Initially, a random initial placement for all the logic blocks is generated. Afterwards, pairs of logic blocks are selected at random as candidates for swapping to improve the cost function. If the swap results in a decrease in the cost function, it is directly allowed; otherwise, it is only allowed with a probability that decreases as the algorithm progresses, thus allowing less worsening swaps after every iteration. A pseudocode for the SA placer is listed in Algorithm 1.1.

Algorithm 1.1 SA generic placer pseudocode

```
S = RandomPlacement()
T = InitialTemperature()
while ExitCriterion() == False do
   /* Outer loop */
   while InnerLoopCriterion () == False do
      /* Inner loop */
      S_new = GenerateViaMove(S)
      ΔC = Cost(S_new) - Cost(S)
      r = random(0,1)
      if r < e^(-ΔC/T) then
         S = S_new
      end if
   end while
   T = UpdateTemp()
end while
```

1.4.4 Timing Analysis

Timing analysis [31] is used to guide placement and routing CAD tools in FPGAs to (1) determine the speed of the placed and routed circuit and (2) estimate the slack of each source–sink connection during routing to identify the critical paths. Timing analysis is usually performed on a directed graph representing the circuit, where the nodes represent LUTs or registers and the edges represent connections.

The minimum required clock period of the circuit can be determined by a breadth-first search through the graph, starting from the primary inputs, to find the arrival time at node i using the following relation:

$$T_{\text{arrival}}(i) = \max_{\forall j \in \text{fanin}(i)} \{T_{\text{arrival}}(j) + \text{delay}(j, i)\}, \qquad (1.1)$$

where delay(j, i) is the delay on the edge between j and i. The required time at node i is calculated by a breadth-first search of the graph, starting from the primary outputs, and using the following relation:

$$T_{\text{required}}(i) = \min_{\forall j \in \text{fanout}(i)} \{T_{\text{required}}(j) + \text{delay}(i, j)\}. \qquad (1.2)$$

Afterwards, the slack on the connection between node i and j is calculated as

$$\text{slack}(i, j) = T_{\text{required}}(j) - T_{\text{arrival}}(i) - \text{delay}(i, j). \qquad (1.3)$$

Connections with a zero slack are critical connections, while those with a positive slack are noncritical ones that can be routed using longer routes.

1.4.5 Routing

The routing phase assigns the available routing resources in the FPGA to the different connections between the logic blocks in the placed design [28]. The objective of a typical routing algorithm is to minimize the delay along the critical path and avoid congestions in the FPGA routing resources. Generally, routing algorithms can be classified into global routers and detailed routers. Global routers consider only the circuit architecture without paying attention to the numbers and types of wires available, whereas detailed routers assign the connections to specific wires in the FPGA.

1.5 VERSATILE PLACE AND ROUTE (VPR) CAD TOOL

VPR is a popular academic placement and routing tool for FPGAs [28]. Almost all the academic works performed on FPGAs is based on the VPR flow. Moreover, VPR is the core for Altera's CAD tool [32, 33]. VPR is usually used in conjunction with T-VPack [18, 27], a timing-driven logic block packing algorithm. VPR consists of two main parts: a placer and router and an area and delay model. These two components interact together to find out the optimum placement and routing that satisfies a set of conditions. This section describes the FPGA architecture supported by VPR as well as gives a quick overview about the tool flow.

1.5.1 VPR Architectural Assumptions

VPR assumes an SRAM-based architecture, where SRAM cells hold the configuration bits for all the pass-transistor MUXs and tristate buffers in the FPGA in both logic and routing resources, as shown in Fig. 1.11a. The SRAMs used are the six-transistor SRAM cells made of minimum size transistors, as shown in Fig. 1.11b. Moreover, an island-style FPGA is assumed by VPR, where the logic clusters are surrounded by routing tracks from all sides. VPR uses an *architecture file* to describe the underlying FPGA architecture used. The architecture file contains information about the logic block size, wire segment length, connection topologies, and other information used by VPR. The use of the architecture file allows VPR to work on a wide range of FPGA architectures. However, there are some

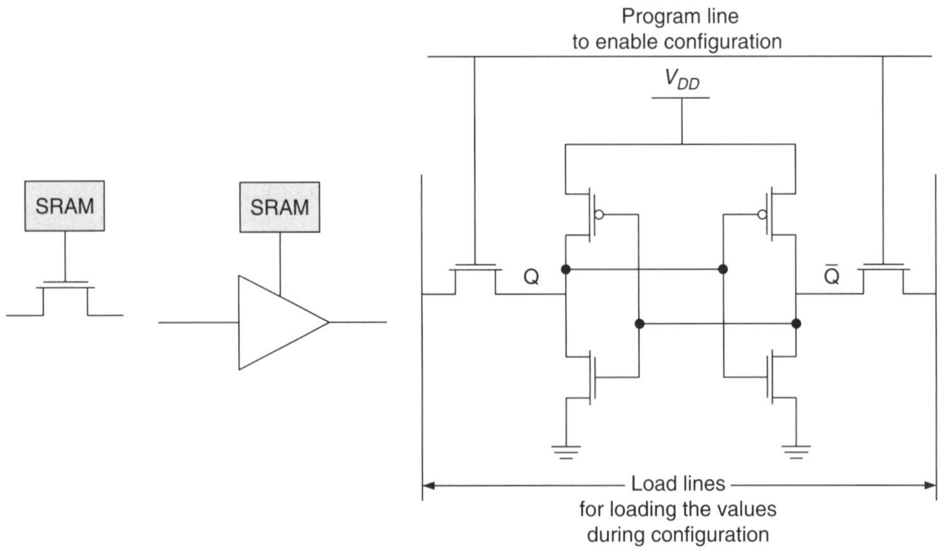

(a) Programmable switches and tristate buffers in VPR FPGA architecture.

(b) Six-transistor SRAM circuit.

■ **FIGURE 1.11** Building blocks of SRAM-programmable FPGAs used by VPR.

general architectural assumptions made by VPR, which are discussed in this section.

1.5.1.1 *VPR Logic Architecture*

VPR targets the hierarchical or cluster-based logic architecture, where every N of the smallest logic element, called *basic logic elements* (BLE), are grouped together to a form a cluster of logic blocks. Each BLE consists of a k-LUT, a D flip-flop (DFF), and a 2:1 MUX, as shown in Fig. 1.12. Such configuration allows both the registered and unregistered versions of the output to be readily available. Local routing resources are used to connect the BLEs inside each logic cluster to each other and to the inputs/outputs of the logic cluster, as shown in Fig. 1.13. As noticed in Fig. 1.13, not all of the BLE inputs are accessible from outside. However, any of the BLE inputs can be connected to any of the BLE outputs or any of the external inputs.

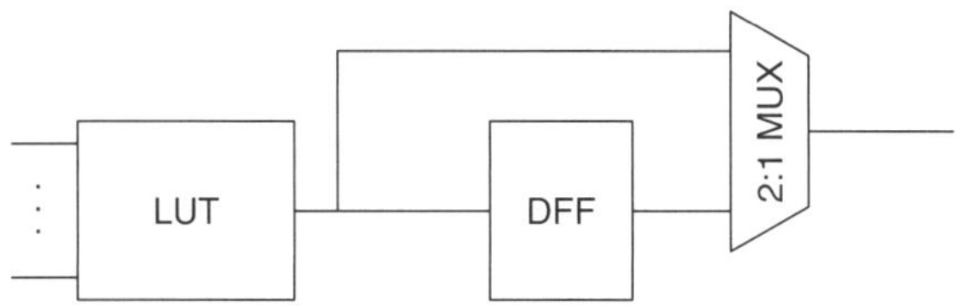

■ **FIGURE 1.12** VPR BLE architecture.

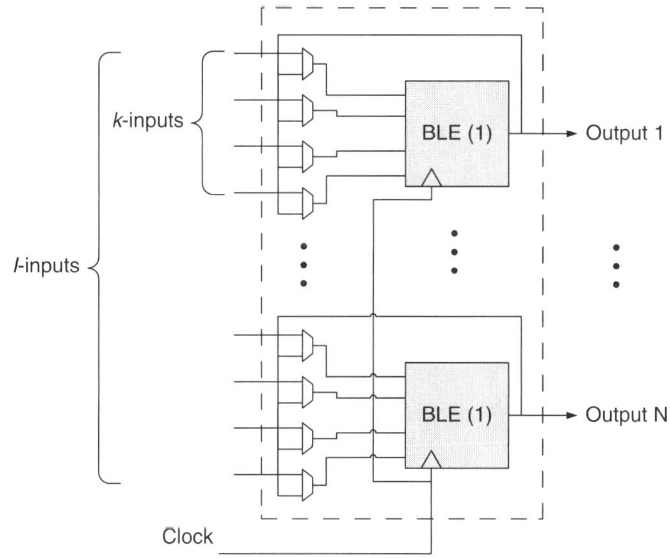

■ **FIGURE 1.13** VPR logic cluster architecture.

A logic cluster is defined in the architecture file by four main parameters: (1) the size of its LUTs, k; (2) the number of BLEs in the cluster, N; (3) the number of external inputs to the cluster, I; and (4) the number of external clock inputs, M_{clk}. It should be noted that VPR assumes minimum-sized transistors are used to implement the LUTs, as a result the capacitances of these transistors are ignored. However, VPR accounts for the capacitance of the internal routing tracks within the logic cluster.

1.5.1.2 *VPR Routing Resources Architecture*

VPR divides the routing resource characterization into three categories: channel, switch block, and wire parameters. The channel information specifies the channel width and the connections between the IO pads and logic blocks from one side and the routing tracks from the other side. The channel width parameters include the width of horizontal (chan_width_x) and vertical (chan_width_y) routing channels, the width of the IO channel (chan_width_io), and the number of IO pads that fit in one row or column of logic clusters (io_rat). The connections between the routing tracks and either the logic blocks or the IO pads are defined by the number of tracks connected to each logic block input ($F_{c,input}$) and output ($F_{c,output}$) and the number of tracks connected to each IO pad ($F_{c,pad}$). As an example, Fig. 1.14 shows a high-level view of a sample VPR FPGA routing model and the values of the parameters used to describe the channel.

Switch blocks are used to provide programmable connectivity between the horizontal and vertical routing tracks, as shown in Fig. 1.14. VPR characterizes switch blocks by their resistance (R), input capacitance (C_{in}), output capacitance (C_{out}), intrinsic delay (T_{del}), connection flexibility (F_s), switch type (whether buffered or not), and switch block topology. The connection flexibility of a switch is defined as the number of connections available for each pin to other pins on the other sides of the switch. Figure 1.11a shows the unbuffered and buffered versions of the switch blocks supported by VPR. Four different topologies of switches can be used within VPR: Disjoint [34], Universal [35], Wilton [36], and Imran [37], as shown in Fig. 1.15.

Finally, VPR describes wire segments by the usage frequency of the segment in the FPGA (segment_frequency), the number of logic clusters spanned by the wire (segment_length), the resistance (R_{metal}) and capacitance (C_{metal}) per unit length, the switch type that connects the wire and logic clusters (opin_switch), and the switch type that connects the wire with other wires (wire_switch).

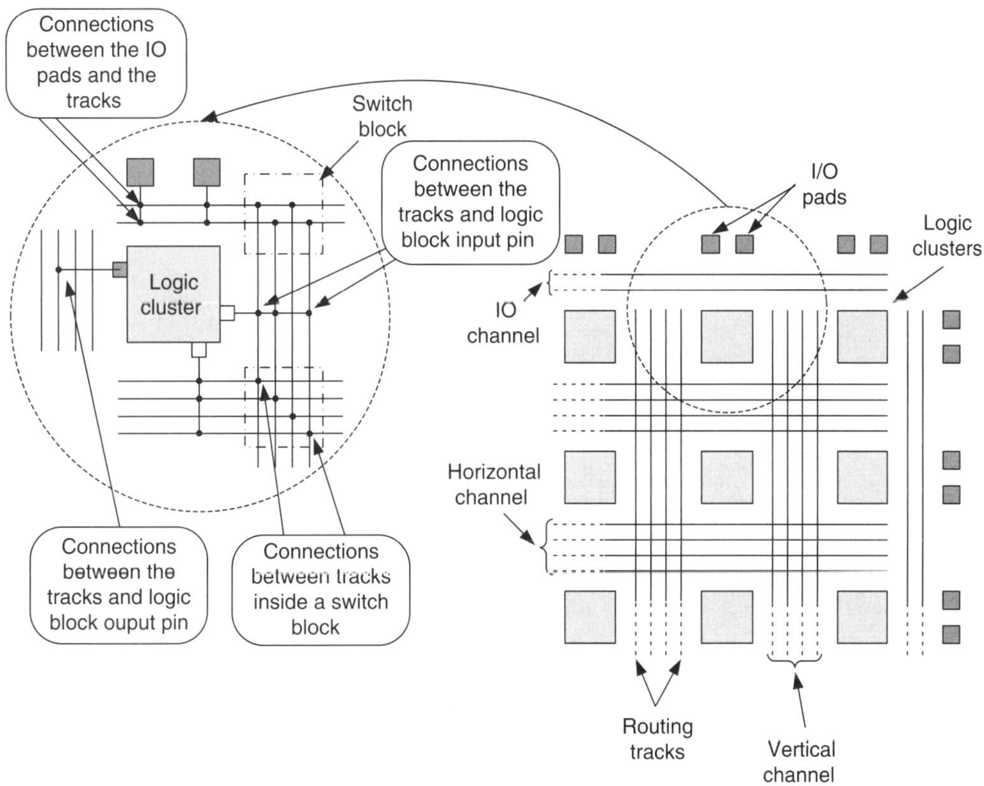

■ **FIGURE 1.14** VPR FPGA routing architecture. chan_width_x = 1, chan_width_y = 1, chan_width_io = 0.5, io_rat = 2, $F_{c,\text{input}} = 3$, $F_{c,\text{output}} = 1$, and $F_{c,\text{pad}} = 2$.

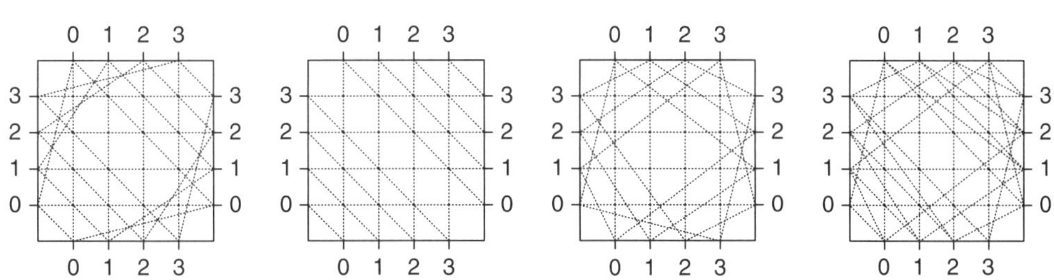

■ **FIGURE 1.15** Switch topologies supported by VPR.

1.5.2 Basic Logic Packing Algorithm: VPack

VPack is a logic packing algorithm that converts an input netlist of LUTs and registers into a netlist of logic clusters. The packing is done in a hierarchical manner in two stages: packing LUTs and registers into BLEs and packing a group of N, or fewer, BLEs into logic clusters. The pseudocode for VPack is listed in Algorithm 1.2.

The first stage of VPack is a pattern matching algorithm that packs a register and an LUT into one BLE when the output of the LUT fansout to *only* one register, as shown in Fig. 1.16.

The second stage packs the BLEs into logic blocks to achieve two objectives: (1) fill the logic clusters to their full capacity N and (2) minimize the number of inputs to each cluster. These two objectives originate from two main goals of the packing: area reduction and improving routability. Packing starts by putting BLEs into the current cluster sequentially in a greedy manner while satisfying the following hard constraints:

1. The number of BLEs must be less than or equal to the cluster size N.
2. The number of externally generated signals, and used inside the cluster, must be less than or equal to the number of inputs to the cluster I.
3. The number of distinct clock signals needed by the cluster must be less than or equal to the number of clock inputs M_{clk}.

A seed BLE is selected for each cluster such that it has the maximum number of inputs among the unclustered BLEs. Other unclustered BLEs, B, are attracted to the cluster, C, in such a way to maximize the Attraction() objective function

$$\text{Attraction}(B) = |\text{Nets}(B) \cap \text{Nets}(C)|, \tag{1.4}$$

Algorithm 1.2 VPack pseudocode [28]

```
Let: UnclusteredBLEs be the set of BLEs not contained in any cluster
- C be the set of BLEs in the current cluster
- LogicClusters be the set of clusters (where each cluster is a set of BLEs)
UnclusteredBLEs = PatternMatchToBLEs (LUTs, Registers)
LogicClusters = NULL
while UnclusteredBLEs != NULL do
   /* More BLEs to cluster */
   C = GetBLEwithMostUsedInputs (UnclusteredBLEs)
   while |C| < N do
      /* Cluster is not full */
      BestBLE = MaxAttractionLegalBLE (C, UnclusteredBLEs)
      if BestBLE == NULL then
         /* No BLE can be added to cluster */
         break
      end if
      UnclusteredBLEs = UnclusteredBLEs - BestBLE
      C = C ∪ BestBLE
   end while
   if |C| < N then
      /* Cluster not full | try hill-climbing */
      while |C| < N do
         BestBLE = MINClusterInputIncreaseBLE (C, UnclusteredBLEs)
         C = C ∪ BestBLE
         UnclusteredBLEs = UnclusteredBLEs - BestBLE
      end while
      if ClusterIsIllegal (C) then
         RestoreToLastLegalState (C, UnclusteredBLEs)
      end if
   end if
   LogicClusters = LogicClusters ∪ C
end while
```

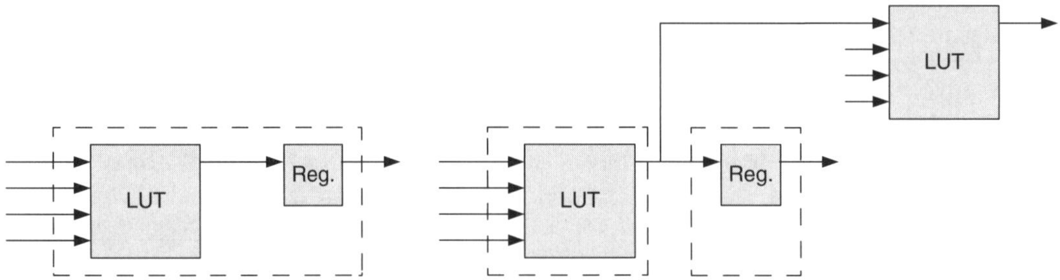

(a) LUT and register packing into one BLE. (b) LUT and register packing into two BLEs.

■ **FIGURE 1.16** Packing LUTs and registers into BLEs [28].

where Nets(x) are the nets connected to BLE (or cluster) x. This process continues until the cluster is filled to its maximum capacity N.

If the cluster does not reach its maximum capacity, but the number of inputs used by the BLEs inside it reaches I, a hill-climbing stage is invoked. In this stage, unclustered BLEs are added to the cluster in such a way to minimize the increase in the number of inputs to the cluster, an example of which is depicted in Fig. 1.17. This is achieved by minimizing the following cost function:

$$\Delta_{\text{cluster inputs}}(B) = |\text{Fanin}(B)| - |\text{Nets}(B) \cap \text{Nets}(C)|. \quad (1.5)$$

It is worth mentioning that the hill-climbing stage allows violating the number of inputs constraints while executing, but does not permit violating the clock inputs constraint. The hill-climbing stage terminates when the cluster size reaches N. If the cluster is infeasible, i.e., its inputs are more than I, the algorithm retracts to the last feasible cluster. Afterwards, VPack selects a new seed BLE and constructs a new cluster.

1.5.3 Timing-Driven Logic Block Packing: T-VPack

T-VPack [18, 27] is a modified version of the VPack algorithm that attempts to minimize the number of intercluster connections along the critical path, besides packing the

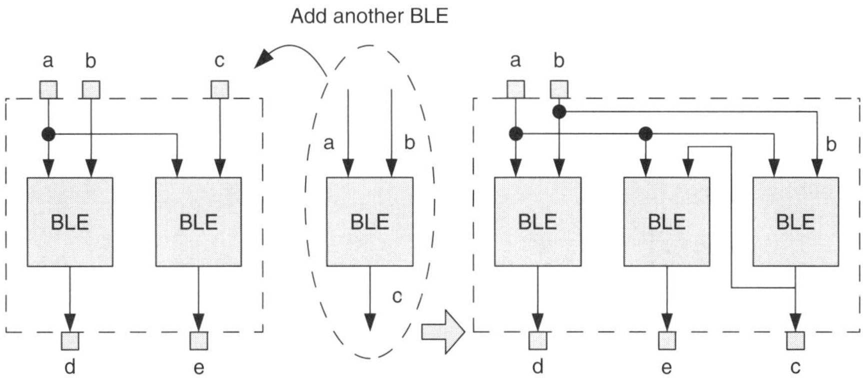

■ **FIGURE 1.17** Adding a BLE to a cluster can decrease the number of used cluster inputs [28].

clusters to their maximum capacity. This achieves speed up along the critical path as local interconnects (intracluster connections) are faster than intercluster interconnects. T-VPack employs a timing analyzer to calculate the slack along the connections in the design and identify the critical path(s). The criticality measure of a connection is calculated as

$$\text{ConnectionCriticality}(i) = 1 - \frac{\text{slack}(i)}{\text{MaxSlack}}, \qquad (1.6)$$

where MaxSlack is the largest slack in the circuit.

In T-VPack, the BLE with the highest criticality, i.e., the BLE connected to the nets with the highest ConnectionCriticality, is selected as the seed BLE for the cluster. Afterwards, BLEs are attracted to the cluster to maximize a modified version of the Attraction() function in Eq. (1.4), given by

$$\text{Attraction}(B) = \lambda \times \text{Criticality}(B) + (1 - \lambda) \times \frac{\text{Nets}(B) \cap \text{Nets}(C)}{\text{MaxNets}}, \qquad (1.7)$$

where *MaxNets* is the maximum number of nets that can connect to any BLE and λ a parameter that controls the trade-off between net sharing and delay minimization. A small value of λ forces T-VPack to focus more on minimizing the number of used inputs to the cluster, and vice versa.

The Criticality(B) of a BLE is given by

$$\begin{aligned}\text{Criticality}(B) = \text{BaseBLECrit}(B) &+ \varepsilon \\ &\times \text{TotalPathsAffected}(B) + \varepsilon^2 \\ &\times D_{\text{source}}(B),\end{aligned} \tag{1.8}$$

where ε is a parameter, $\text{D}_{\text{source}}(B)$ is the BLE distance or level from the source of the path, TotalPathsAffected(B) is the total number of critical paths connecting primary inputs/outputs or registers inputs/outputs and B, and BaseBLECrit(B) is the base criticality of BLE B and is evaluated as: (i) the maximum ConnectionCriticality(i) of all the connection to/from B while selecting a seed BLE, or (ii) the maximum ConnectionCriticality(i) of all the connections joining BLE B to the other BLEs in the cluster currently being packed.

1.5.4 Placement: VPR

VPR models the FPGA as a block array of logic clusters bounded by routing tracks, as shown in Fig. 1.18. SA is used as the optimization algorithm for placement in VPR using an adaptive annealing schedule to adapt to the current placement at any time instant.

The initial temperature is selected from the basic features of the circuit. Assume that the total number of logic clusters in the design is N_{clusters}. After the initial random placement is evaluated, N_{clusters} pairwise swaps are performed and the initial temperature is calculated as 20 times the standard deviation of the cost of the different N_{clusters} combinations evaluated. Moreover, the

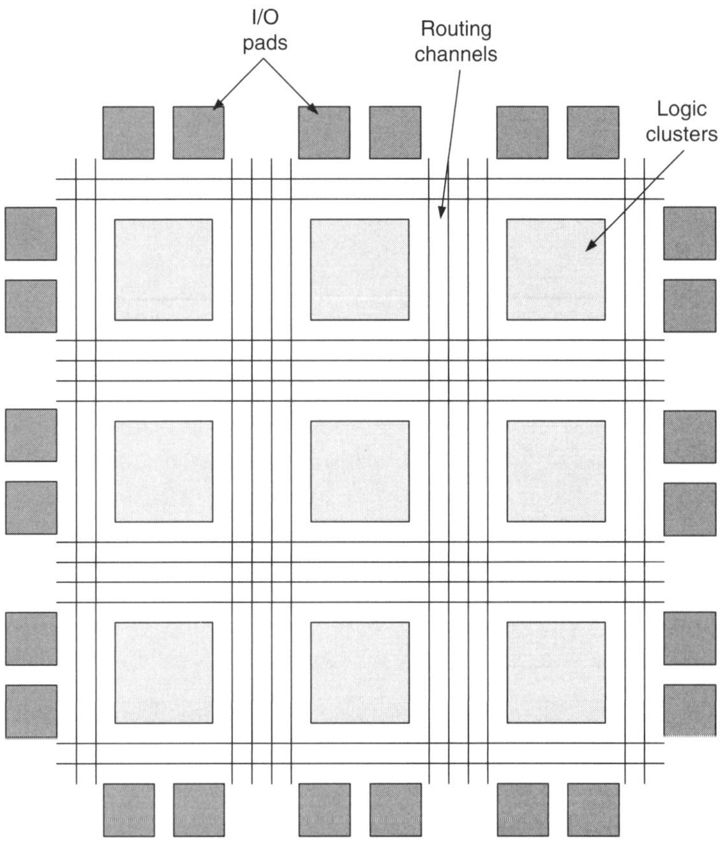

■ **FIGURE 1.18** FPGA model assumed by the VPR placer.

number of inner moves performed at each temperature is evaluated as

$$\text{MovesPerTemperature} = \text{InnerNum} \times N_{\text{clusters}}^{4/3}, \tag{1.9}$$

where InnerNum is a constant and usually set to 10.

Another feature of the adaptive SA algorithm used in VPR is the way the temperature is updated. In conventional SA, almost all the moves are accepted at high temperatures, while at low temperatures only improving moves

Table 1.1 VPR Temperature Update Schedule [28]

α	γ
$0.96 < \alpha$	0.5
$0.8 < \alpha \leq 0.96$	0.9
$0.15 < \alpha \leq 0.8$	0.95
$\alpha \leq 0.15$	0.8

are accepted. In the adaptive SA [28], the cooling scheme tries to prolong the time spent in these cost-improving temperatures (medium and low temperatures) at the expense of possibly cost-worsening temperatures (high temperatures) using the following temperature update relationship

$$T_{\text{new}} = \gamma \times T_{\text{old}}, \tag{1.10}$$

where γ is evaluated with respect to the percentage of moves accepted (α), according to Table 1.1.

1.5.5 **Routing: VPR**

VPR incorporates two different routing algorithms: a routability-driven router and a timing-driven router.

1.5.5.1 ***Routability-Driven Router***

The VPR routability-driven router is based on the *Pathfinder* algorithm [38]. The Pathfinder algorithm repeatedly rips up and reroutes every net in the circuit during each routing iteration until all congestions are removed. Initially, all nets are routed to minimize the delay, even if this results in congestion. Afterwards, routing iterations are applied to overused routing resources to resolve such congestions. In VPR, the cost of using a routing resource n when it is reached by connecting it to routing resource m is given by

$$\mathrm{Cost}(n) = b(n) \times h(n) \times p(n) + \mathrm{BendCost}(n, m), \quad (1.11)$$

where $b(n)$, $h(n)$, and $p(n)$ are the base cost, historical congestion, and present congestion, respectively. $b(n)$ is set to the delay of n, delay(n). $h(n)$ is incremented after each routing iteration in which n is overused. $p(n)$ is set to "1" if routing the current net through n will not result in congestion and increases with the amount of overuse of n. The BendCost(n, m) is used to penalize bends in global routing to improve the detailed routability.

Timing-Driven Router

The timing-driven router in VPR is based on the Pathfinder, but timing information is considered during every routing iteration. Elmore delay models are used to calculate the delays, and hence, timing information in the circuit. To include timing information, the cost of including a node n in a net's routing is given by

$$\begin{aligned}\mathrm{Cost}(n) = {} & \mathrm{Crit}(i,j) \times \mathrm{delay}(n, \mathrm{topology}) \\ & + [1 - \mathrm{Crit}(i,j)] \times b(n) \times h(n) \times p(n), \quad (1.12)\end{aligned}$$

where a connection criticality Crit(i,j) is given by

$$\mathrm{Crit}(i,j) = \max\left\{\left[\mathrm{MaxCrit} - \frac{\mathrm{slack}(i,j)}{D_{\max}}\right]^{\eta}, 0\right\}, \quad (1.13)$$

where $D_{\max}$ is the critical path delay and η and MaxCrit are parameters that control how a connection's slack impacts the congestion-delay trade-off in the cost function.

Chapter 2

Power Dissipation in Modern FPGAs

The tremendous growth of the semiconductor industry in the past few decades is fueled by the aggressive scaling of the semiconductor technology following Moore's law. As a result, the industry witnessed an exponential increase in the chip speed and functional density with a significant decrease in power dissipation and cost per function [39]. However, as complementary metal oxide semiconductor (CMOS) devices enter the nanometer regime, leakage current is becoming one of the main hurdles to Moore's law. According to Moore, the key challenge for continuing process scaling in the nanometer era is leakage power reduction [40]. Thus, circuit designers and CAD engineers have to work hand in hand with device designers to deliver high-performance and low-power systems for future CMOS devices. In this chapter, the power dissipation problem is discussed in the VLSI industry in general and in FPGAs in particular.

2.1 CMOS TECHNOLOGY SCALING TRENDS AND POWER DISSIPATION IN VLSI CIRCUITS

The main driving forces that govern the CMOS technology scaling trend are the overall circuit requirements: the maximum power dissipation, the required chip speed, and the needed functional density. The overall device requirements such as the maximum MOSFET leakage current, minimum MOSFET drive current, and desired transistor size are determined to meet the overall circuit requirements. Similarly, the choices for MOSFET scaling and design, including the choice of physical gate length L_g and equivalent oxide thickness of the gate dielectric t_{ox}, and so forth, are made to meet the overall device requirements. Figure 2.1 depicts the scaling trend for the CMOS feature size across several technology generations as well as some future predictions according to the semiconductor road map published by the International Technology Roadmap for Semiconductors (ITRS) [41].

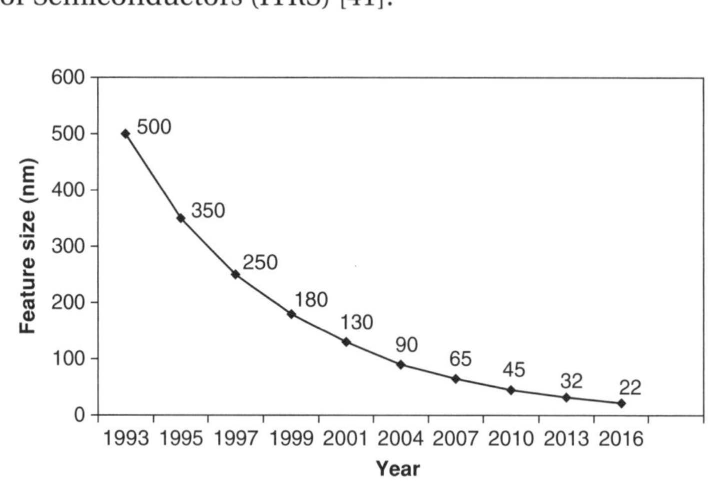

■ **FIGURE 2.1** Gate length scaling of CMOS technologies [41].

There are two common types of scaling trends in the CMOS process: constant field scaling and constant voltage scaling. Constant field scaling yields the largest reduction in the power-delay product of a single transistor. However, it requires a reduction in the power supply voltage as the minimum feature size is decreased. Constant voltage scaling does not suffer from this problem, therefore, it provides voltage compatibility with older circuit technologies. The disadvantage of constant voltage scaling is the electric field increases as the minimum feature length is reduced, resulting in velocity saturation, mobility degradation, increased leakage currents, and lower breakdown voltages. Hence, the constant field scaling is the most widely used scaling approach in the CMOS industry. Table 2.1 summarizes the constant field scaling in the CMOS process.

Table 2.1 Constant Field Scaling of the CMOS Process

Parameter	Symbol	Constant Field Scaling
Gate length	L	$1/\alpha$
Gate width	W	$1/\alpha$
Field	ε	1
Oxide thickness	t_{ox}	$1/\alpha$
Substrate doping	N_a	α
Gate capacitance	C_G	$1/\alpha$
Oxide capacitance	C_{ox}	α
Circuit delay	t_d	$1/\alpha$
Power dissipation	P_d	$1/\alpha^2$
Area	A	1
Power density	P/A	1

To maintain the switching speed improvement of the scaled CMOS devices, the threshold voltage V_{TH} of the devices is also scaled down to maintain a constant device overdrive. However, decreasing V_{TH} results in an exponential increase in the subthreshold leakage current,

$$I_D \propto 10^{\frac{V_{GS}-V_{TH}+\eta V_{DS}}{S}}, \tag{2.1}$$

where $S = \frac{nkT}{q} \ln 10$. Moreover, as the technology is scaled down, the oxide thickness t_{ox} is also scaled down, as shown in Table 2.1. The scaling down of t_{ox} results in an exponential increase in the gate oxide leakage current.

As a result of the continuous scaling of V_{TH} and t_{ox}, the contribution of the total leakage power to the total chip power dissipation is increasing notably. The contribution of leakage power is expected to exceed 50% of the total chip power by the 65 nm CMOS process [41], as shown in Fig. 2.2.

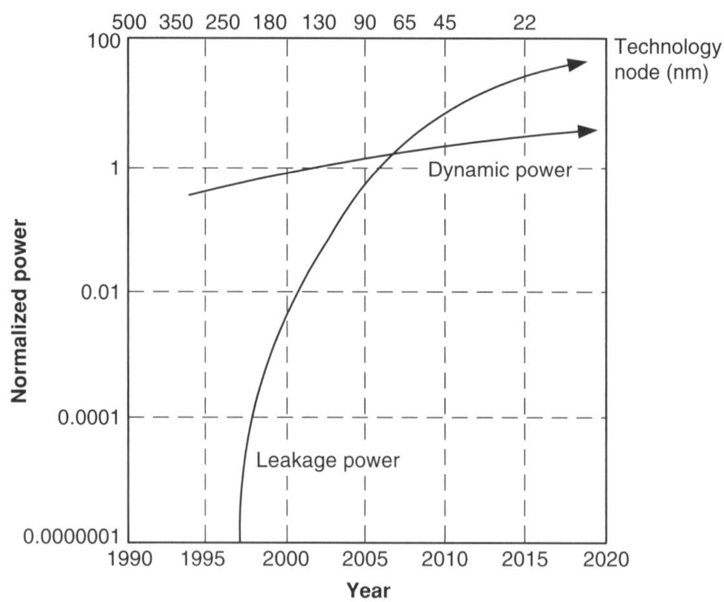

■ **FIGURE 2.2** Leakage power contribution to the total chip power [41].

2.2 DYNAMIC POWER IN FPGAs

FPGAs provide reconfigurability by using redundant logic and switches inside the chip. As a result, FPGA design dynamic power dissipation is much larger than their application-specific integrated circuit (ASIC) counterparts. In a study by Kuan and Rose [42], the authors performed a quantitative study to compare the dynamic power dissipation in FPGAs to that of ASICs, and the results are listed in Table 2.2.

The results in Table 2.2 suggest that, on average, FPGAs consume 14× more dynamic power than ASICs when the circuits contain only logic. However, when the design uses some of the hard blocks inside the FPGA, e.g., memory blocks and multipliers, the dynamic power gap is reduced between FPGAs and ASICs, with multipliers being the main factor in reducing FPGA power dissipation. The main reason for this reduction in the power gap is due to the fact that using the hard macros inside the FPGAs means fewer logic resources are used, hence, less power is being dissipated. As a result, it can be concluded that FPGAs are less efficient in terms of power dissipation when compared to ASICs. In order for FPGAs to be able to compete with ASICs, extensive work is still needed to reduce FPGA dynamic power dissipation.

2.3 LEAKAGE POWER IN FPGAs

2.3.1 CMOS Device Leakage Mechanisms

There are six short-channel leakage current mechanisms in CMOS devices. Figure 2.3 summarizes the leakage current types that affect state-of-the-art CMOS devices [43]. I_1 is the reverse-bias pn junction leakage; I_2 is the subthreshold leakage; I_3 is the oxide tunneling current; I_4 is the gate current due to hot-carrier injection; I_5 is the gate-induced drain leakage; and I_6 is the channel punchthrough current. Currents I_2, I_5, and I_6 are OFF-state leakage currents, while I_1 and I_3 occur in both ON and OFF states.

Table 2.2 Dynamic Power Consumption Ratio (FPGA/ASIC) [42]

Circuit	Logic Only	Logic and DSP	Logic and Memory	Logic, Memory, and DSP
booth	26	-	-	-
rs_cncodcr	52	-	-	-
cordic18	6.3	-	-	-
cordic8	5.7	-	-	-
desarea	27	-	-	-
des_perf	9.3	-	-	-
fir_restruct	9.6	-	-	-
mac1	19	-	-	-
aesl92	12	-	-	-
tir3	12	7.5	-	-
diffeq	15	12	-	-
diffeq2	16	12	-	-
molecular	15	16	-	-
rs_dccodcrl	13	16	-	-
rs_decoder2	11	11	-	-
atm	-	-	15	-
aes	-	-	13	-
aes_inv	-	-	12	-
ethernet	-	-	16	-
serialproc	-	-	16	-
fir24	-	-	-	5.3
pipe5proc	-	-	-	8.2
raytracer	-	-	-	8.3
Geomean	14	12	14	7.1

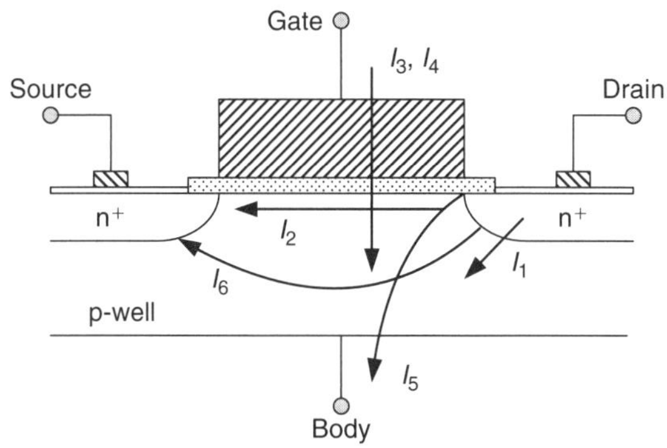

■ **FIGURE 2.3** Leakage current mechanisms of deep submicron devices [43].

I_4 can occur in the OFF state, but more typically occurs during the transistor transition [43]. The main sources for leakage power dissipation in current CMOS technolo gies are the subthreshold leakage and gate oxide leakage currents.

There are two main components for the reverse-bias pn junction leakage I_1: minority carrier diffusion/drift near the edge of the depletion region and electron-hole pair generation in the depletion region of the reverse-biased junction. I_2 flows between the source and drain in a MOSFET when the gate voltage is below V_{th}. I_3 occurs by electrons tunneling from the substrate to the gate and also from the gate to the substrate through the gate oxide layer. I_4 occurs due to electrons or holes gaining sufficient energy from the applied electric field to cross the interface potential barrier and enter into the oxide layer. I_5 is due to the high field effect in the drain junction of the MOSFET. Because of the proximity of the drain and the source, the depletion regions at the drain-substrate and source-substrate junctions extend into the channel. Channel length reduction and increase in the reverse bias across the junctions push the junctions nearer to each other until they almost merge, thus leading to the punchthrough current I_6.

Of these six different leakage current mechanisms experienced by current CMOS devices, subthreshold and gate leakage currents are the most dominant leakage currents. Furthermore, the contribution of subthreshold leakage current to the total leakage power is much higher than that of gate leakage current, especially at above room temperature operating conditions. The contribution of gate leakage current to the leakage power dissipation is expected to increase significantly with the technology scaling, unless high-k materials are introduced in the CMOS fabrication industry [41].

2.3.2 Current Situation of Leakage Power in Nanometer FPGAs

For FPGAs to support reconfigurability, more transistors are used than those used in an ASIC design that performs the same functionality. Consequently, leakage power dissipation in FPGAs is higher than that in their ASIC counterpart. It was reported in a study by Kuon and Rose [42] that on average, the leakage power dissipation in FPGA designs is almost 5.4 times that of their ASIC counterparts under worst-case operating conditions. The excess leakage power dissipated in FPGAs is mainly due to the programming logic that is not present in ASIC designs.

A study of the leakage power dissipation in a 90-nm CMOS FPGA was performed by Tuan and Lai [44], the results of which are summarized in Table 2.3. By comparing the average leakage power dissipation of a typical 90-nm CMOS FPGA at 25°C and 85°C, it can be seen that the average leakage power increases by four times. Moreover, the results in the first column are for a utilization of 75%; hence, the leakage power dissipation for a 1000-CLB FPGA would be in the range of 4.2 mW. If these FPGAs are to be used in a wireless mobile application, which has a typical leakage current of 300 μA, then the maximum number of CLBs that can be used would be 86 CLBs for the 25°C and 20 CLBs for the 85°C.

Table 2.3 FPGA Leakage Power for Typical Designs and Design-Dependent Variations [44]

T	Typical P_{LEAK} (avg. Input Data; $U_{CLB} = 75\%$)	Best-Case Input Data	Worst-Case Input Data
25°C	4.25 μW/CLB	−12.8%	+13.0%
85°C	18.9 μW/CLB	−31.1%	+26.8%

In addition, the dependence of leakage on the input data increases significantly with the temperature. This can be deduced from Table 2.3 as the variation due to the worst and best case input vectors change from ±13% at 25°C to approximately ±28% at 85°C. Furthermore, in another experiment conducted by Tuan and Lai [44], it was found out that for a 50% CLB utilization, 56% of the leakage power was consumed in the unused part of the FPGA. Hence, in future FPGAs, these unused parts have to be turned down to reduce this big portion of leakage power dissipation.

Chapter 3

Power Estimation in FPGAs

With power dissipation posing as an important factor in the design phase of FPGAs, power estimation and analysis techniques have turned out to be huge challenges for both FPGA vendors and designers. Power characterization is an important step in designing power efficient FPGA architectures and FPGA applications. FPGA designers need a method to quantify the power advantage of the architectural design decisions without having to go through fabrication. Moreover, FPGA users need to check the power efficiency of the several possible implementations without actually going through the lengthy design phase.

Power models for FPGAs need to consider both components of power dissipation: dynamic power and leakage power. To further improve the accuracy of the power model, all the subcomponents of both dynamic power (switching, short circuit, and glitch power) and leakage power (subthreshold and gate leakage power) need to be accounted for. In addition, other factors that affect power dissipation, including spatial correlation and input dependency of leakage power, should be considered, especially in the subnanometer regime since their impact is highlighted as the CMOS minimum feature size is scaled down.

3.1 INTRODUCTION

Modern CMOS processes suffer from two dominant sources of power dissipation: dynamic and leakage power. Dynamic power dissipation can be divided into switching, glitch, and short-circuit power dissipation, whereas leakage power dissipation can be further divided into subthreshold leakage and gate leakage power dissipation. Historically, CMOS circuits were dominated by dynamic power dissipation; however, by the 65 nm CMOS process, leakage power is expected to dominate the total power dissipation, as explained in Chapter 2. In addition, as the

CMOS process is further scaled down, gate leakage power is expected to surpass the subthreshold power dissipation, especially at lower operating temperatures, as predicted by the semiconductor road map issued by the ITRS [41], unless high-k materials are used to implement the device gates.

All sources of power dissipation in CMOS circuits exhibit significant state dependency. To develop an accurate power model, accurate information about signal probabilities should be made available. Several studies in the literature have addressed this problem and a complete survey was presented in the study by Najm [45].

The switching component of the dynamic power dissipation is expressed as

$$Power_{\mathrm{dyn}} = \frac{1}{2} \times f_{\mathrm{clk}} \times V_{\mathrm{DD}} \times V_{\mathrm{swing}} \times \sum_{i=1}^{n} C_i \times \alpha_i, \quad (3.1)$$

where f_{clk} is the circuit clock frequency, V_{DD} is the circuit supply voltage, V_{swing} is the swing voltage, C_i is the capacitance of the ith node in the circuit, and α_i is a measure of the number of transitions per clock cycle experienced by node i. As a result, the problem of switching power estimation resolves to find the capacitance and estimating the number of transitions at every node.

Glitching power occurs because of the spurious transitions at some circuit nodes due to unbalanced path delays. Glitches are hazardous transitions that do not contribute to the circuit functionality. Consequently, glitching power results in an increase in the number of transitions at every circuit node that is susceptible to glitches. Hence, the impact of glitching power can be modeled by adding a factor to the transitions estimate α in Eq. (3.1).

Finally, short-circuit power dissipation is the power dissipated due the presence of a direct current path from the power supply to the ground during the rise and fall times of each transition. Hence, short-circuit power is a function of the rise and fall times and the load capacitance. Several research projects have been directed to provide an accurate estimate of short-circuit power dissipation [46–49]. However, the simplest method to account for short-circuit power dissipation is to set it as a percentage of the dynamic power dissipation, usually 10% [50].

In most of the power estimation techniques in the literature, spatial independence among the signals was assumed. Under the spatial independence assumption, all the signals are assumed independent even though they might share a common parent gate; hence, the effect of reconvergent paths is ignored. This assumption is used to significantly reduce the power calculation runtime; however, it results in significant inaccuracies in the power estimation. It was reported in the study by Schneider and Krishnamoorthy [51] that the relative error in switching activities estimation under the spatial independence assumption in VLSI designs can reach 50%.

3.2 POWER ESTIMATION IN VLSI: AN OVERVIEW

The power estimation is defined as the problem of evaluating the *average* power dissipation in a digital circuit [45]. Power estimation techniques fall into two main categories: simulation-based or probabilistic-based approaches. In this section, a brief overview of the two power estimation techniques is presented.

3.2.1 Simulation-Based Power Estimation Techniques

In simulation-based techniques, a random sequence of input vectors is generated and used to simulate the circuit to estimate the power dissipation. The first approaches developed were based on the use of SPICE simulations to simulate the whole circuit using a long sequence of input vectors [52, 53]. However, the use of such methods in today's VLSI industry is impractical, especially with the huge levels of integration achieved, that it might take days if not weeks to simulate a complete chip using SPICE. Moreover, these methods are significantly pattern dependent due to the use of a random sequence of input vectors. If an intelligent method is used to select the input sequence, these methods would provide the most accurate power estimation.

Several simplifications of these methodologies were proposed to reduce the computational complexity of power estimation [54–57]. The methods still rely on simulations, but instead of using SPICE simulations, other levels of circuit simulations were performed including switch-level and logic-based simulations. These methodologies trade accuracy for faster runtime. To perform these simulations, the power supply and ground are assumed constant. However, these methods still suffer from pattern dependency since there is a need to generate a long sequence of input vectors to achieve the required accuracy. Although these simulations are more efficient than SPICE simulations, they are somewhat impractical to use for large circuits, especially if a long input sequence is used to increase the method accuracy.

To solve the pattern dependency problem of simulation-based power estimation methods, several statistical methods had been proposed [58–63]. These statistical methods aim to quantify two parameters: the *length of the input sequence* and the *stopping criteria* for simulation, required to achieve a predefined power estimate accuracy.

The earliest of these studies focused on the use of Monte Carlo simulations to estimate the total average power [58]. A random sequence of N input vectors was independently generated and used to simulate the circuit. Let $\overline{p}$ and s be the average and standard deviation of the power measured over a time period T and P_{av} is the average power. Hence, the error in the average power estimated can be expressed with a confidence of $(1-\alpha) \times 100\%$ as

$$\frac{|\overline{p} - P_{\mathrm{av}}|}{\overline{p}} < \frac{t_{\alpha/2}s}{\overline{p}\sqrt{N}}, \tag{3.2}$$

where $t_{\alpha/2}$ is generated from the t-distribution with $(N-1)$ degrees of freedom [58]. Hence, to tolerate a percentage error of ϵ, the required length of the input sequence is expressed as [58]

$$N \geq \left(\frac{t_{\alpha/2}s}{\epsilon\overline{p}}\right)^2. \tag{3.3}$$

An extension of this work was proposed by Xakellis and Najm [59] to provide an estimate of the average power dissipation in each gate instead of the whole circuit.

A disadvantage of the use of Eq. (3.3) is that the value of N cannot be estimated before simulation. In the study by Hill and Kang [60], they proposed a different formulation of the required length on input vectors a priori to simulation. A single-rising-transition approximation was adopted for circuits that do not experience glitches, and the value of N for an error of ϵ and confidence of $(1-\alpha)$ is approximated by [60]

$$N \approx \frac{z^2_{1-\alpha/2}}{\epsilon^2}, \tag{3.4}$$

where $z^2_{1-\alpha/2}$ is the $100 \times (1-\alpha/2)$th percentile of the standard normal distribution. For circuits with glitches or large logic depths, N for an error of ϵ and confidence of $(1-\alpha)$ is approximated by [60]:

$$N \approx \frac{4z^2_{1-\alpha/2}}{49\epsilon^2} \times (t+1)^2, \tag{3.5}$$

where t is the maximum number of transitions that the circuit can experience per input vector.

Another disadvantage with Eq. (3.3) is the large number of input vectors needed to achieve the required accuracy, since it depends on the square of the sample variance. Moreover, as the resulting power estimate deviates from the normal distribution, the simulation might terminate early, thus compromising the accuracy of the results. In the studies by Marculescu et al [61] and Liu and Papaefthymiou [63], solutions to these two issues were proposed using a Markov chain to generate the input sequence. The resulting sequences are more compact than the ones used in the study by Burch et al [58] and provide a reduction in the simulation time by orders of magnitude while keeping the estimated average power within 5%.

Another statistical method proposed in the literature to provide the needed length for the input sequence is based on the least square estimation methods [62]. The authors viewed the estimation problem of the input sequence as an approximation problem and explored the use of sequential least square and recursive least square to solve the problem of finding the input sequence that has minimum variance and without making any probabilistic assumptions about the data. It was reported by Murugavel et al [62] that least square algorithms need a much smaller number of iterations, i.e., smaller input sequence, to provide a close estimate of the average power dissipation to that reported in the study by Burch et al [58].

3.2.2 **Probabilistic-Based Power Estimation Techniques**

Probabilistic power estimation techniques have been proposed to solve the problem of pattern dependency of simulation-based approaches. In these techniques, signal probabilities are propagated through the circuit starting from the primary inputs until the outputs are reached. These estimation techniques require circuit models for probability propagation for every gate in the library.

The first ever probabilistic propagation model was proposed in the study by Cirit [64]. In this model, a *zero-delay assumption* was considered, under which the delay of all logic gates and routing resources was assumed zero. The switching activity of node x was defined as the probability that a transition occurs at x. The *transition probability* of x, $P_{\mathrm{t}}(x)$ is calculated according to

$$P_{\mathrm{t}}(x) = 2 \times P_{\mathrm{s}}(x) \times P_{\mathrm{s}}(\overline{x}) = 2 \times P_{\mathrm{s}}(x) \times \left[1 - P_{\mathrm{s}}(x)\right], \tag{3.6}$$

where $P_{\mathrm{s}}(x)$ and $P_{\mathrm{s}}(\overline{x})$ are the probabilities $x = 1$ and $x = 0$, respectively. In adopting Eq. (3.6), the authors assume that the values of the same signal in two consecutive clock cycles are independent, which is referred to as *temporal independence*. Moreover, the signal probabilities at the primary inputs are propagated into the circuit while assuming that all internal signals are independent. This assumption is referred to as *spatial independence*. Furthermore, Cirit [64] ignores glitching power since a zero delay model was adopted.

Najm [65], proposed the use of the transition density to represent the signal probabilities more accurately than the simple transition probability in Eq. (3.6). The transition density is defined as the average number of transitions per second at a node in the circuit. The transition density at node x, $D(x)$, is given by

$$D(x) = \lim_{T \to \infty} \frac{n_x(T)}{T}, \tag{3.7}$$

where $n_x(T)$ is the number of transitions within time T. Najm [65] formulated the relationship between the transition density and the transition probability by

$$D(x) \geq \frac{P_t(x)}{T_{\mathrm{c}}}, \tag{3.8}$$

where T_{c} is the clock cycle. Hence, the transition probability will always be less than the transition density, thus underestimating the power dissipation. The transition density at node y with a set of inputs $x_0, x_1, \ldots, x_n$ is given by the following set of relationships:

$$D(y) = \sum_{i=1}^{n} P\left(\frac{\partial y}{\partial x_i}\right) D(x_i), \tag{3.9}$$

$$\frac{\partial y}{\partial x_i} \triangleq y|_{x_i=1} \oplus y|_{x_i=0}, \tag{3.10}$$

where $\frac{\partial y}{\partial x_i}$ is the Boolean difference of y with respect to its ith input and $\oplus$ denotes exclusive OR operation. To evaluate the Boolean difference at each node, the probabilities at each node need to be propagated through the whole circuit. It should be noted that the use of (3.9) only provides a better estimate for the number of transitions than the transition density given in Eq. (3.6) and this model still suffers from both spatial and temporal independence assumptions.

In the study by Ghosh et al [66], the authors proposed the use of binary decision diagrams (BDDs) to account for spatial and temporal correlations. The regular Boolean function of any logic gate stores the steady-state value of the output given the inputs. However, BDDs are used to store the final value as well as the intermediate states, provided that circuit delays are available beforehand. As a result, such a probabilistic model can predict the signal probabilities at each circuit node under spatial and temporal correlations. However, this technique is computationally expensive and only practical for moderate-sized circuits. In addition, a BDD is required for every logic gate. If some gates have a large number of intermediate states, then this technique might become impractical even for medium-sized circuits.

Several other research projects have been proposed in the literature to handle spatial and temporal correlations [67–72].

3.3 COMMERCIAL FPGA POWER ESTIMATION TECHNIQUES

Commercial FPGA vendors offer a variety of power estimation techniques for customers. There are two categories of commercial FPGA power estimation techniques: device-specific spreadsheets [73–75] and CAD-based power estimation techniques [74, 76, 77].

3.3.1 Spreadsheet Power Estimation Tools

FPGA power spreadsheets analyze both the leakage and dynamic power dissipation components in the FPGA. This method of power estimation is usually used in the early stages of the design process to give a quick estimate of the power dissipation of the design [73–75]. In spreadsheet-based power estimators, the users must provide the clock frequency of the design and the toggle percentage for the logic blocks. Consequently, this method gives a rough approximation of power and requires designers to thoroughly understand the switching activity inside their circuits.

For dynamic power computation, designers provide the average switching frequency α for all the logic blocks, or for each module in the design. Coefficients for adjusting the dynamic power calculation are provided in the device data sheet. The total dynamic power is calculated according to

$$P_{\text{dyn}} = K \times f_{\text{clk}} \times V_{\text{DD}} \times \sum_{\text{all\_components}} \alpha_i C_i V_{\text{swing},i}, \quad (3.11)$$

where K is the coefficient used in adjusting the power estimate for each device family. The value of K is measured through empirical experiments performed by the FPGA vendor to encompass the impact of both glitching and short-circuit power components. The values of both K and the node capacitance are usually provided as an average value for the combination of family/die/package being used. The only variables that are design specific are the clock frequency and the average switching probability of the design. As a result, the average dynamic power estimated using this approach is very crude and should only be used as a guideline on which family/die/package to be used for a certain design.

In modeling leakage power dissipation, power spreadsheets list the leakage power of each FPGA component $P_{\text{leak\_per\_component}}$. For example, the spreadsheet lists the average leakage power per logic block, sequential flip-flop or latch, and IO cell. The total leakage is calculated by summing the leakage power for each component used in the design according to

$$P_{\text{leak}} = \sum_{\text{all\_components}} P_{\text{leak\_per\_component}} \times N_{\text{component}}. \tag{3.12}$$

Similarly, the value of $P_{\text{leak\_per\_component}}$ is provided for each combination of FPGA family/die/package. It should be noted that the leakage power estimated using this approach is device specific rather than design specific.

The total estimated power is the summation of the dynamic and leakage power evaluated using Eqs. (3.11) and (3.12), respectively.

3.3.2 **CAD Power Estimation Tools**

Commercial CAD power estimation tools provide different variants of power estimators with different accuracies. The most accurate power estimation tool provided relies on cycle accurate simulations to capture the switching at each node in the design [74, 76, 77]. The user specifies a simulation test bench, and the design is simulated using logic simulators. The logic simulator records all the transitions that occur on every net in the design.

Afterwards, these transitions are read using the power estimator, and using the value of the capacitance at each node, the total dynamic power of the design can be easily estimated using Eq. (3.1). The accuracy of the power estimation can be further improved by using postlayout capacitance for power calculation. Although this power estimation technique provides a better estimate for power dissipation than spreadsheets, the runtime of this technique is very long and the power estimate accuracy is dependent on the length of the test vector used, similar to simulation-based power estimation techniques discussed above.

Another method of power estimation used by commercial FPGA power estimators is vectorless power estimation. This approach is mainly a probabilistic power estimation methodology, where the probabilities at every net are propagated through the circuit to estimate the total dynamic power estimation. Again, postlayout node capacitances can be used to improve the power estimation accuracy. Commercial vectorless power estimation tools do not take into account glitching power or spatial correlation between signals, thus suffering from reduced accuracy compared to the simulation-based approach.

FPGA vendors provide another simulation-based power estimation methodology that does not depend on cycle accurate simulation. A random input vector is used to simulate the circuit while counting the transitions that occur at every node. Hence, glitching power is ignored in that case, while spatial correlation between the different signals is taken into consideration. Since cycle accurate simulation is not used in that case, simple logic simulators can be used, thus reducing the runtime of that approach, while keeping the accuracy of the power estimated somewhere between the first two approaches mentioned above.

Finally, for leakage power estimation, most commercial FPGA CAD power estimators use an approach similar to that used in power spreadsheets. An extra parameter that is used to increase the accuracy of leakage power estimation is accounting for the power dissipated in the unused resources of the FPGA. This is achieved by multiplying the number of the unused resources on the die by the average leakage power of the unused resources. Hence, it should be noted that none of the commercial FPGA CAD tools accounts for the state dependency of leakage power dissipation.

3.4 A SURVEY OF FPGA POWER ESTIMATION TECHNIQUES

Power modeling in FPGAs did not receive wide attention in the literature, especially for generic FPGA architectures. Several works targeted specific FPGA architectures [44, 78–82]. These works are only applicable to one architecture because they depend on the specific architecture details to extract the power dissipation. Moreover, all these works targeted only dynamic power estimation in FPGAs, except for the study by Tuan and Lai [44] that provided an insight into leakage power dissipation in a 90 nm CMOS commercial FPGA. On the other hand, power dissipation in general architecture FPGA has been targeted in a limited number of research projects [83–87]. However, all these research projects did not provide an analytical methodology for estimating the effect of spatial correlation on the total power dissipation.

In the study by Kusse and Rabaey [78], a Xilinx XC4003A™ FPGA was used to study the power breakdown inside the FPGA. The power reported was the actual power measure recorded from the physical FPGA itself. Weiß et al [79] introduced a technology-dependent empirical correction factor for dynamic power estimation in the Xilinx Virtex™ FPGA. The factor introduced was used to adjust the switching activity estimated through regular probabilistic analysis similar to the study by Najm [65]. The power dissipation in the FPGA under consideration was physically measured for different benchmarks, and the power dissipation was calculated in a similar manner to that of the study by Najm [65] for the same benchmarks. The ratio between the measured and estimated power values was calculated as the correction factor for power estimation. The main reason for using that correction factor is to account for the factors that affect power dissipation, including temporal and spatial correlations, which are not captured in the probabilistic calculations presented in the study by Najm [65]. It should be noted that this power estimation methodology is strongly dependent on the benchmarks used to compute the value of the correction factor as different types of benchmarks will result in different values of the factor. It should be noted that both the studies by Kusse and Rabaey [78] and Weiß et al [79] do not account for leakage power dissipation in the FPGA under consideration.

A model for dynamic power dissipation in Xilinx Virtex-II™ FPGAs was described in the study by Shang et al [80]. The switching activity was estimated using logic simulation of some practical input vectors obtained from the users. The different node capacitances were evaluated from post–silicon capacitance extraction techniques. An extension of this work was presented by Degalahal and Tuan [81], where the node capacitances were evaluated using simple RC models for the Xilinx Spartan-3™ FPGA. The methodology is still simulation-based since it relies on logic-level simulation to find the node activities.

3.4.1 Linear Regression-Based Power Modeling

Anderson and Najm [82] proposed a power model for FPGAs that predicts accurate switching activities in Xilinx FPGAs using curve fitting theories and empirical formulae. The main goal of this work is to account for glitches in the transitions activity used in dynamic power calculations. A prediction function was proposed to calculate the change in transitions activity to account for glitching in the form of

$$PR_i = \alpha \times \mathrm{GEN}_i + \beta \times \mathrm{GEN}_i^2 + \gamma \times \mathrm{PROP}_i + \upsilon \times \mathrm{PROP}_i^2 + \nu \times D_i + \xi \times D_i^2 + \eta \times \mathrm{PROP}_i \times \mathrm{GEN}_i + \iota \times \mathrm{GEN}_i \times D_i + \rho \times \mathrm{PROP}_i \times D_i + \phi, \quad (3.13)$$

where α, β, γ, υ, ν, ξ, η, ι, ρ, ϕ are scalar constants, GEN_i is a parameter used to quantify the number of glitches generated at node i, PROP_i represents the amount of glitch propagation at node i, and D_i is a parameter to represent the depth of node i. GEN_i is formulated as

$$\mathrm{GEN}_i = \min_{x_i \in \mathrm{inputs}(y)} \{|PL_y| - |PL_{x_i}|\}, \quad (3.14)$$

where PL_y is the set of different paths from the primary inputs to node y and is given by

$$PL_y = \bigcup_{x_i \in \mathrm{inputs}(y)} \{p + 1 | p \in PL_{x_i}\}. \quad (3.15)$$

The authors used a commercial Xilinx Virtex-II™ PRO FPGA, and using a predefined input sequence and a commercial simulation tool, they generated the number of transitions experienced by every gate per clock cycle in several FPGA benchmarks. Afterwards, using curve fitting theories, the constants in Eq. (3.13) were evaluated.

Although the method proposed in their study [82] tries to formulate the impact of glitches on dynamic power estimation in FPGAs, it has several drawbacks. First, the method is architecture dependent and cannot be readily used to estimate dynamic power dissipation in other FPGA architectures than the one used. Moreover, to apply this method on a new architecture, the linear regression model needs to be trained; hence, there is a need for a reference power estimator to train the linear regression model.

In addition, the model accuracy is strongly dependent on the types of circuits used to evaluate the curve fitting parameters, which makes it very susceptible to errors for the different circuit types.

3.4.2 Probabilistic FPGA Power Models

The power models that target general architectures consider dynamic, leakage, and short-circuit power dissipation in FPGAs. The first power model for generic FPGA architectures was proposed by Poon et al [83]. This power model is analytic in nature, making it very easy to implement with fast runtime. However, for this model to have such fast runtime, several approximations were assumed by the authors. First, for dynamic power estimation, they [83] assume spatial and temporal independencies among the internal design signals, as well as ignore the impact of glitching power. Second, the leakage power was calculated across all the transistors in the circuit while considering the V_{GS} to be half the threshold voltage V_{TH}, thus, significantly reducing the accuracy of the leakage power estimation. Moreover, the state dependency of leakage power was not considered in that model, which has a significant impact on FPGAs built using current nanometer CMOS technologies, as explained in the study by Tuan and Lai [44]. However, the power model proposed by Poon et al [83] had several advantages, including fast runtime and ability to be integrated within VPR. As a result, the power model became the de facto for power estimation for academic research.

3.4.3 Look-up Table–Based FPGA Power Models

A study of the leakage power dissipation in the CMOS 90 nm Xilinx Virtex-II™ FPGA was presented by Tuan and Lai [44]. The leakage power modeling was performed using look-up tables (LUTs) of HSpice simulations. The leakage power for every input vector measured, obtained using

HSpice simulations, was recorded and used to study the impact of the state dependency and utilization on FPGA leakage power dissipation. The study showed that the input state dependency can vary leakage power in modern FPGAs by approximately 60%. Moreover, a breakdown of leakage power dissipation in the different parts of the FPGA was provided. It should be noted that this study ignored the effect of signal correlations on leakage power. Moreover, Tuan and Lai [44] did not evaluate the total leakage power dissipation in the whole FPGA, only evaluated the leakage power per logic block.

Another FPGA power model was presented for generic FPGAs [84–86]. Leakage power dissipation was calculated through the use of LUTs, which do not consider the state dependency of leakage power. Moreover, the power model depended on logic simulation to estimate the switching activities of the different circuit nodes. Although this method can achieve high accuracy, its computational cost is quite high. The authors tried to limit the execution time by limiting the number of input vectors to 2000, irrespective of the circuit size and its number of inputs. However, this approach sacrifices the accuracy of the algorithm significantly as explained in the study by Chou and Roy [88].

The input dependency of leakage power dissipation in generic FPGAs was first addressed by Kumar and Anis [87]. The authors proposed a leakage power model based on the BSIM4 models while accounting for the state dependencies. However, spatial correlation among internal signals was not addressed. Moreover, the authors used some empirical constants in the power formulation obtained using curve fitting, thus, rendering the power model technology dependent. Finally, the temperature dependence of power dissipation in FPGAs was studied by Lui et al [89]. The authors tried to estimate a factor that captures the dependence of total power dissipation on the temperature using empirical experimentation.

3.5 A COMPLETE ANALYTICAL FPGA POWER MODEL UNDER SPATIAL CORRELATION

Hassan et al proposed a complete analytical power model for FPGAs that accounts for spatial correlation between the circuit signals [90]. In this section, the details of this power model will be presented.

3.5.1 Spatial Correlation and Signal Probability Calculations

The signal probability can be computed at the output of each logic block through simple probabilistic calculations to evaluate the probability of the output of the logic block being high. In all the available probabilistic power models for FPGAs, the inputs to any specific logic block are assumed independent, the spatial independence assumption. This assumption is made to simplify the probability calculation. However, the spatial independence assumption reduces the accuracy of any analytical power model by overestimating the signal probabilities, as will be explained later. As an example, for the small circuit shown in Fig. 3.1, assuming that the probability of $A=1$ is 0.5, dynamic power estimators operating under the spatial independence assumption will calculate the probability of B being high as

$$P(B=1) = P(A=1) \times P(\overline{A}=1) = 0.5 \times 0.5 = 0.25, \quad (3.16)$$

thus resulting in a transition probability of 0.375 ($2 \times 0.25 \times 0.75$), according to Eq. (3.6). However, by clear inspection of Fig. 3.1, this circuit suffers from a reconvergent path.

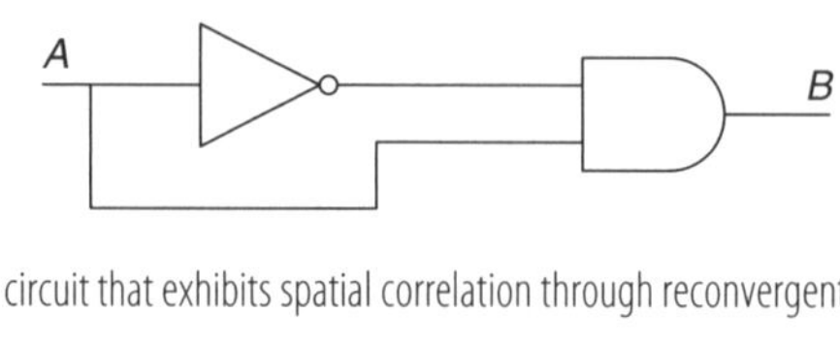

■ **FIGURE 3.1** A circuit that exhibits spatial correlation through reconvergent paths.

Both A and $\overline{A}$ can never be "1" simultaneously; hence, the probability of B being high should be zero, assuming the zero delay model. It should be noted that such a structure is common in FPGA circuits, which will be depicted later in Table 3.4.

As noticed from the previous example, the spatial independence assumption can significantly affect the dynamic power calculations. If spatial correlation is to be taken into consideration, conditional probability should be used to evaluate the probability of all signals. As an example, considering the circuit in Fig. 3.1, the probability of B being high is formulated as

$$P(B=1) = P(A=1|A=1) \times P(\overline{A}=1|A=1) \times P(A=1) + P(A=1|A=0) \times P(\overline{A}=1|A=0) \times P(A=0), \quad (3.17)$$

which resolves to zero. From the above example, it can be deduced that the spatial independence assumption can significantly affect the accuracy of power estimation in VLSI circuits in general. Moreover, by inspecting Eq. (3.17), it can be noticed that spatial correlation will cancel one or more of the conditional probabilities listed. Hence, the impact of spatial independence will always be toward overestimating the power dissipation by overestimating the signal probabilities. Thus, the spatial independence assumption costs the designers in terms of overdesign to account for the overestimated power dissipation.

Hassan et al proposed a methodology to calculate the signal probabilities under spatial correlation [90]. To consider spatial correlation for power calculations, such reconvergent paths as the one shown in Fig. 3.1 need to be identified, as discussed in Section 3.5.2, and their signal probabilities corrected accordingly, as explained in Section 3.5.3. The proposed methodology is explained in the following two sections.

3.5.2 Exploration Phase: Locating Spatial Correlation

Spatial correlation among signals in VLSI circuits occurs whenever two signals are correlated. Correlations arise when two or more signals share a common driver or a common parent logic block (x) and are connected as inputs to another logic block (y), i.e., reconvergent paths. If the circuit is converted to a cyclic graph with the gates as the nodes and the signal wires as the edges, the connections between x and y form a cycle with two paths, as shown in Fig. 3.2. Hence, the detection of signals that might exhibit spatial correlation resolves to identify possible cycles in the circuit. For example, in Fig. 3.2, a cycle would be detected that goes through $A \rightarrow \bar{a} \rightarrow \bar{A} \rightarrow b \rightarrow A$.

In the study by Hassan et al [90], a depth-first search algorithm is used to identify such loops in the design. The algorithm starts with the circuit primary inputs i, and depth-first search is used to navigate through all the logic blocks that share a path with each primary input. Whenever a logic block j is visited, it is marked with the name of the primary input used in this search. When a logic block j gets visited twice, this means that there are two paths from the current primary input i to logic block j. Afterwards, the two different paths are recorded as cycles that might result in spatial correlation among the inputs to logic block j. By using the depth-first search algorithm, the complexity of the exploration phase gets significantly reduced. A pseudocode for the exploration phase is shown in Fig. 3.1.

As a result of the Algorithm 3.1, every logic block that experiences spatial correlation among its inputs will have all the paths that contribute to the reconvergent paths recorded. However, these paths can be very long, especially in circuits with long logic depth. As a result, a cleanup stage

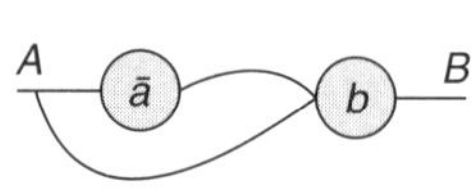

FIGURE 3.2 A graph representation of the circuit in Fig. 3.1.

Algorithm 3.1 The exploration phase pseudocode used to identify reconvergent paths in a circuit [90]

```
Function: explore
for all primary inputs i do
   for all blocks j connected to i do
      depth_first(j,i)
   end for
end for
return
Function: depth_first(j,i)
if j has been visited before by i then
   a cycle is found
   record the path i → j
else
   label j as visited by i
end if
for each block k connected to j do
   depth_first(k,i)
end for
return
```

is performed on these paths to remove the redundancy in these paths. For example, in Fig. 3.2, the two paths recorded for B would be $A \rightarrow \overline{A}$ and A. If A is not a primary input, then the two paths will also include logic blocks that generate the inputs to A in the circuit and so on until the primary inputs. Hence, the cleanup stage deletes all the nodes in the path that are common and only keeps the fanout stem of the reconvergent path, which is A in this example.

3.5.3 Signal Probabilities Calculation Algorithm under Spatial Correlation

Once all the cycles in the circuit that contribute to spatial correlation are identified and recorded, the algorithm starts correcting the signal probabilities for all the logic blocks that have correlated inputs. As a first step, all the logic blocks are sorted in a topological order according to their connections. In this ordering, the primary inputs come in first, followed by those logic blocks that only have primary inputs as their inputs, and so on. This ordering is essential because it is the same order at which the signal probabilities and hence transition densities are calculated according to Eq. (3.9).

In the second step, all the signal probabilities in the design are calculated based on the spatial independence assumption among the circuit signals. Processing the logic blocks according to the topological ordering performed earlier ensures that whenever a logic block is processed, all its inputs have already been processed and signal probabilities have been calculated for them.

In the third step, the logic blocks that have cycles are examined. For each logic block that has cycles, the number of different fanout stems for the cycles are recorded. As an example, in Fig. 3.3, logic block E will have two cycles with A and B being the fanout stem of the two cycles. The first cycle has A as the first path and $A \rightarrow C$ as the second path. The second cycle has $B \rightarrow C$ as the first path and $B \rightarrow D$ as the second path. It should be noted that these two cycles are not independent, they both share C, thus making the calculation of the signal probabilities more complex.

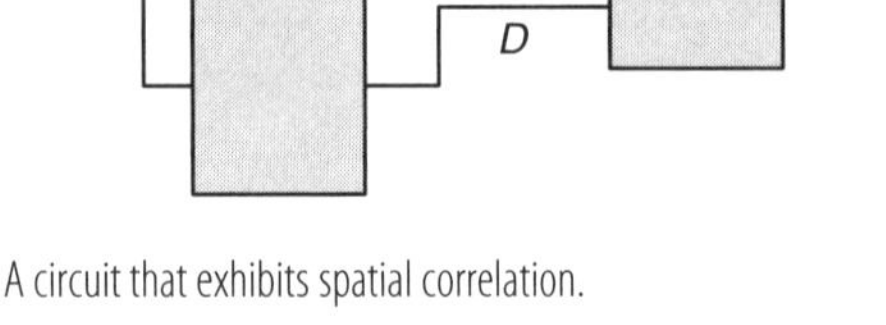

■ **FIGURE 3.3** A circuit that exhibits spatial correlation.

If the number of fanout stems for logic block i is n, then for the conditional probability calculations, there are 2^n different conditional probabilities to calculate for every input to the logic block that experiences reconvergent paths. For the circuit in Fig. 3.3, the conditional probabilities that need to be calculated are

$$\begin{aligned}
&P(C = 1|A = 0\&B = 0), \quad P(D = 1|A = 0\&B = 0)\\
&P(C = 1|A = 0\&B = 1), \quad P(D = 1|A = 0\&B = 1)\\
&P(C = 1|A = 1\&B = 0), \quad P(D = 1|A = 1\&B = 0)\\
&P(C = 1|A = 1\&B = 1), \quad P(D = 1|A = 1\&B = 1)
\end{aligned} \tag{3.18}$$

The algorithm starts by assuming the first combination for the fanout stems, which is $A = 0$ and $B = 0$ in this case, and then propagates through all the cycle paths recorded for logic block E and evaluates their probabilities. This phase is stopped when all the probabilities of the inputs to block E are evaluated. Afterwards, the conditional probability for block i is calculated and multiplied by the probability of occurrence of the tested input combination, i.e., $P(A = 0\&B = 0)$ in the example in Fig. 3.3. This process continues until all the 2^n combinations are processed. The probability of logic block E output being high under spatial correlation will be the summation of the probabilities evaluated for each input combination for block E according to Eq. (3.18). This process continues on until all of the logic blocks in the design are processed. The importance of the cleanup phase mentioned in Section 3.5.2 is that it reduces the number of probability calculations to a greater extent by getting rid of the common paths. A pseudocode for the algorithm is listed in Algorithm 3.2.

The function Adjust_Prob_Fanout_Stems(j) in Algorithm 3.2 converts the integer j to its binary equivalent and adjusts the probabilities of the fanout stems accordingly. As an example, for the circuit in Fig. 3.3, if $j = 2$, then $P(A) = 1$ and $P(B) = 0$. Find_Prob_Fanout_Stems(j) finds the probability of the fanout stems combination given

Algorithm 3.2 Probabilities calculation under spatial correlation algorithm [90]

```
Order_Logic_Blocks()
Calc_Prob_Under_Independence()
for each block i with cycles do
  n = Find_Num_Fanout_Stems(i)
  prob_i = 0
  for j=0 : j = 2^n do
    Adjust_Prob_Fanout_Stems(j)
    prob_fanout_stems = Find_Prob_Fanout_Stems(j)
    prob_i = prob_i + Find_Prob(i) × prob_fanout_stems
  end for
end for
```

by j, e.g., when $j = 2$, prob_fanout_stems $= P(A = 1) \times P(B = 0)$, assuming that A and B are independent. If A and B are not dependent, then the probability of $P(A) = 1$ and $P(B) = 0$ was calculated by the algorithm when it processed A and B. Find_Prob(i) evaluates the probability of block i for the current input combination of the fanout stems. This is performed by evaluating the probabilities of all the logic blocks in the paths contributing to the cycles connected to block i. It should be noted that similar algorithms had been used in the literature for fault detection in VLSI circuits [91–94].

By inspecting Algorithm 3.2, it can be deduced that the complexity of the algorithm is $O(m \times 2^n \times k)$, where m is the number of logic blocks with cycles, n is the number of fanout stems that any logic block can have, and k is the maximum number of cycles that any logic has. Signal probabilities under spatial correlation depend on the maximum number of cycles handled. If all the cycles at the input of any logic block are handled, then the algorithm would have the highest accuracy at the expense of the increased complexity and execution time. Hence, having a maximum for the number of cycles to be considered by the algorithm would result in slightly less accurate value for the probabilities but with a faster runtime.

The proposed algorithm was executed first for all the FPGA benchmarks listed in Table 3.4 while considering all the cycles present in the design. Afterwards, a maximum limit on the number of fanout stems for each node to be considered was set and the algorithm was executed several times for different maximum values. The paths that are rejected are the paths in which their children have the lowest probabilities. For example, in Fig. 3.3, if one of the paths is to be rejected, if the $P(C = 1) < P(B = 1) < P(A = 1)$, then the paths that have C in them are rejected. This decision is taken because they will have the least impact on the final probability [59]. It was found that when the number of fanout stems handled by the algorithm was limited to five, the accuracy of the signal probabilities calculated, when compared to the first case, had an error below 4%, while the algorithm execution time got reduced significantly when compared to the case with all the cycles. Hence, in the study by Hassan et al [90], the number of fanout stems handled is limited to five to reduce the algorithm complexity while achieving the best accuracy. The results of this experiment are presented later on in Section 3.5.6.

3.5.4 Power Calculations Due to Glitches

Glitches occur in VLSI circuits due to the difference in the arrival times of the inputs to any logic block, e.g., both A and C have different arrival times as inputs of E in Fig. 3.3. To identify the logic blocks that might generate glitches, the postlayout arrival times of all the design signals are extracted using VPR [28]. VPR takes as input the capacitances and resistances of the different wire segments in the FPGA fabric under consideration. Using this information, VPR calculates the arrival times for every signal at every circuit node using simple Elmore delay calculations. There are two conditions needed for glitch generation: (1) the differences in the arrival times should be larger than the intrinsic delay of the logic cell and (2) the logic implemented results in a glitch. Moreover, it should be noted that glitches are filtered out of the circuit through retiming elements such as latches and buffers.

The proposed algorithm for glitch probability calculations consists of three phases: glitch generation, glitch propagation, and glitch termination. Starting with the logic cells connected to the primary inputs, and parsing the circuit in a depth-first strategy, when conditions (1) and (2) are satisfied, glitches are generated at the output of the logic cell. For instance, in Fig. 3.3, if a glitch at E occurs when A switches to "1" and C to "0", then the probability that such a glitch occurs is

$$P_g(E) = P(C = 0|A = 1) \times P(A = 1). \tag{3.19}$$

It should be noted that the algorithm for calculating the signal probabilities under spatial correlation calculates the conditional probability in Eq. (3.19) if they are correlated; otherwise, Eq. (3.19) resolves to $P(C = 0) \times P(A = 1)$.

When a glitch from the output of one cell is fed to the input of the next cell, that glitch propagates only if the logic function of the second cell allows it to. The probability of that a certain glitch will propagate is equal to the probability of the glitch multiplied by the conditional probabilities of the inputs needed to propagate the glitch. The proposed glitch propagation algorithm keeps on parsing the circuit by the depth-first search until the probability of glitch propagation is less than 0.01, at which point the glitch is dropped, *glitch termination*. It should be noted that no new probabilities are calculated by the glitch processing algorithm.

3.5.5 **Signal Probabilities and Power Dissipation**

In this section, the effect of signal probabilities on the components of power dissipation is discussed. Moreover, the method used in the study by Hassan et al [90] to model both dynamic and static power dissipation is presented.

3.5.5.1 ***Dynamic Power Dissipation***

Using the transition density in calculating the dynamic power dissipation using Eq. (3.1) results in [65]

$$P_{\text{dyn}} = \frac{1}{2} \times f_{\text{clk}} \times V_{\text{DD}}^2 \sum_{i=1}^{n} C_i D_i, \tag{3.20}$$

where D_i is the transition density of node i. The transition density is calculated from the signal probabilities using Eqs. (3.9) and (3.10) [65]. From the above equations, it is shown that the calculation of the transition density depends strongly on the proper calculation of the signal probabilities. Consequently, spatial correlation directly affects the accuracy of the transition density calculation. The transition density at each node is calculated efficiently by simple propagation algorithms that depend on the signal probabilities at each node [65].

The above discussion is valid for combinational logic; however, for sequential circuits, some approximations are made. In the study by Tsui et al [95], iterations were used to calculate the output probability of sequential feedback loops. Initially, the input and output probabilities are set to the same value. Then, by performing several probability calculation iterations, the output probability is adjusted. Tsui et al [95] also demonstrated that the transition probability of the feedback loop is within 5% compared to the exact transition probability value, provided that a sufficient number of iterations are performed. In the study by Hassan et al [90], the same methodology proposed by Tsui et al [95] and used in the study by Poon et al [50] is used to calculate the transition density at the output of sequential feedback loops. It should be noted that this methodology has been selected due its ease of implementation, rather than the quality of its results.

The capacitances used in Eq. (3.20) are extracted from commercial CMOS processes using the postlayout capacitance extractor available in Cadence tools. A small fabric is designed using the fully custom design flow and the layout of the circuit was performed together with the routing tracks and multiplexers. Afterwards, Cadence is used to extract the resistances and capacitances of all the routing tracks with different lengths in our FPGA architecture. The chosen frequency in the study by Hassan et al [90] to calculate the power dissipation is 600 MHz. This value was chosen because it corresponds to the maximum clock frequency at which state-of-the-art FPGAs operate [9, 10].

3.5.5.2 *Leakage Power Dissipation*

In FPGAs, logic functions and routing resources are implemented using pass-transistor-based multiplexers, as shown in Fig. 3.4. In the logic resource LUTs, the inputs (S0–S3) are connected to SRAM cells, while the controls (C0–C1) are connected to the inputs of the LUT. However, in the FPGA routing resources, the inputs are connected to the signals to be routed and the controls are connected to SRAM cells that control the routing switch.

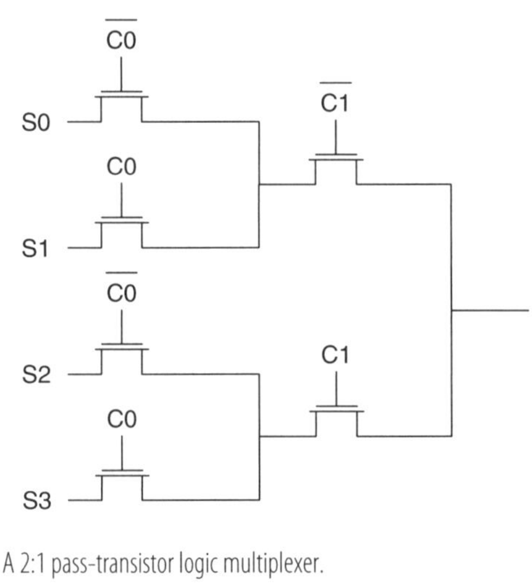

■ **FIGURE 3.4** A 2:1 pass-transistor logic multiplexer.

Anderson et al [96] demonstrated the dependence of leakage power dissipation in the pass-transistor-based multiplexers on the input vector. For a 90 nm CMOS process, the leakage power dissipation in the pass-transistor multiplexer in Fig. 3.4 can vary by 14 times depending on the input combination [96]. Two main factors affect the threshold voltage of the pass transistors and, hence, leakage power dissipation in these multiplexers: body effect and drain-induced barrier lowering (DIBL). The effect of body bias on V_{TH} is formulated as

$$V_{TH} = V_{TH0} + \gamma\left(\sqrt{\Phi_s - V_{BS}} - \sqrt{\Phi_s}\right), \tag{3.21}$$

where V_{TH0} is the ideal V_{TH} at zero V_{BS}, γ is the body bias coefficient, and Φ_s is the surface potential. Having a negative V_{BS} would result in increasing the subthreshold voltage, which in turn will reduce the subthreshold leakage current. It should be noted that CMOS devices in pass-transistor multiplexers will never experience a positive V_{BS}. Pass transistors with logic 0 or opposite signal polarity at both terminals will not experience body effect because their V_{BS} would be zero. However, those devices with logic 1 at both the terminals will experience subthreshold leakage current reduction due to body effect because their $|V_{BS}|$ would be maximum (either V_{DD} or $V_{DD} - V_{TH}$).

In nanometer CMOS devices, the DIBL effect causes the threshold voltage to be a function of the drain-source voltage. Applying a large drain-source voltage to the CMOS device results in decreasing the subthreshold voltage, hence increasing the subthreshold current. For minimum-sized 90 nm NMOS devices, V_{TH} can vary by almost 25% and leakage current by 4.5 times due to a difference in V_{DS} equal to the supply voltage.

Pass-transistor multiplexers used in FPGAs can experience four different values of V_{DS}. The transistors in the first and last stages of the multiplexer are the only ones that can experience the worst-case V_{DS} of V_{DD}. The middle stages

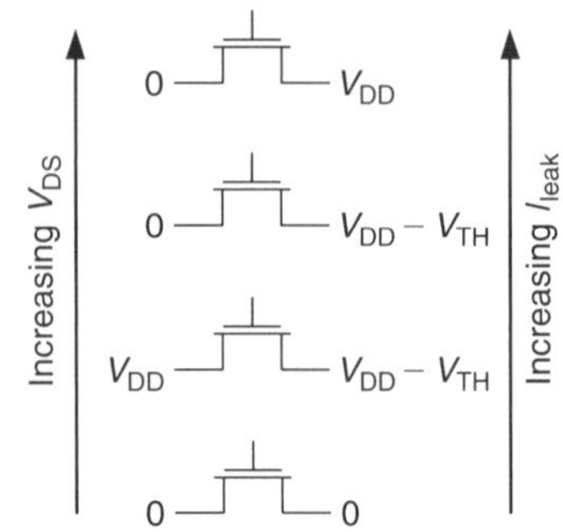

■ **FIGURE 3.5** DIBL impact on subthreshold leakage in FPGA pass-transistor devices.

can experience a maximum of $V_{DD} - V_{TH}$ because of the weak "1" passed by the NMOS pass transistors. Figure 3.5 shows the four different values of V_{DS} that the pass transistors can experience in FPGAs and the impact on leakage current. Since the signal probability is an indication of the probability that a certain signal is high, then a more accurate leakage power model needs to take into account the different signal states. This fact was used by Kumar and Anis [87] to develop the first FPGA leakage power model that considers state dependency; however, spatial independence was assumed.

In the study by Hassan et al [90], the pass-transistor multiplexer in Fig. 3.4 is simulated using HSpice using all the possible input combinations, and the resulting leakage power dissipation is recorded in each case. The leakage values are recorded in a LUT and used in the power modeling technique. The total leakage power dissipation in any multiplexer in the design is the sum of the leakage power dissipation for a certain input combination (P_{leak_i}) multiplied by the probability of occurrence of this combination (P_i),

$$P_{\text{leak}} = \sum_{i=0}^{l} P_{\text{leak}_i} \times P_i, \tag{3.22}$$

where l is the total number of input combinations. Using Eq. (3.22), the proposed power model will take into consideration the state dependency of subthreshold leakage power under spatial correlation if the probabilities are computed under spatial correlation using the algorithm presented in Section 3.5.3.

3.5.5.3 *Gate Leakage Power Dissipation*

Under the predictions of ITRS [41], the contribution of gate leakage is expected to increase significantly compared to subthreshold leakage power in future technology nodes. Unlike subthreshold leakage, gate leakage is available in both the ON and OFF states of the CMOS devices. The value of gate leakage is again a strong function of both V_{GS} and V_{DS}. Large values of V_{GS} and small values of V_{DS} result in a larger gate leakage current. Hence, an accurate power model for future FPGAs should consider the state dependency, including spatial correlation, of gate leakage power dissipation.

In the study by Hassan et al [90], the values of the gate leakage of all the basic circuit elements that are used in FPGAs are evaluated using HSpice simulations. The values of the gate leakage current under all the input combinations are recorded in a LUT and used to evaluate the gate leakage power in a similar manner to Eq. (3.22).

3.5.6 **Results and Discussion**

The proposed power estimation methodology under spatial correlation is implemented and integrated into the VPR CAD tool [28]. To evaluate the performance of the proposed algorithm for signal probability estimations under spatial correlation, several experiments were performed. First, to test the accuracy of the algorithm in evaluating the signal probabilities for the different signals in the design, several FPGA benchmarks are simulated using a logic simulator under the zero-delay assumption.

A pseudorandom input vector is applied to the inputs of each benchmark and the signal probabilities of the circuit internal nodes are recorded. The length of the input vector used is 10^5, which is proven to result in small inaccuracies [88]. To quantify the accuracy of the proposed algorithm, the following metrics are used. The average relative error of the signal switching activity is used as a metric of the algorithm accuracy in estimating the switching activity [51]

$$e = \frac{1}{\#\ \text{signals}} \sum_{\text{signal } i} \left| \frac{\alpha_{i,\text{alg}} - \alpha_{i,\text{sim}}}{\alpha_{i,\text{sim}}} \right|, \quad (3.23)$$

where $\alpha_{i,\text{alg}}$ and $\alpha_{i,\text{sim}}$ are the switching activities estimated by the proposed algorithm and the input vector simulation method, respectively. We also define the maximum and minimum relative errors of the signal switching activity as follows:

$$e_{\text{min}} = \min_{\text{signal } i} \left| \frac{\alpha_{i,\text{alg}} - \alpha_{i,\text{sim}}}{\alpha_{i,\text{sim}}} \right|, \quad (3.24)$$

$$e_{\text{max}} = \max_{\text{signal } i} \left| \frac{\alpha_{i,\text{alg}} - \alpha_{i,\text{sim}}}{\alpha_{i,\text{sim}}} \right|. \quad (3.25)$$

The switching activities evaluated from the input simulations are then compared to those evaluated from the proposed algorithm and the study by Poon et al [83]. The resulting relative errors are reported in Table 3.1. The benchmarks marked with a gray background are those with a combinational section while the others are datapath circuits. It can be noticed from Table 3.1 that the proposed algorithm manages to capture the correlation between the internal signals of the design even though only five cycles were included in the switching activity estimation. The average e_{max} resulting from the proposed algorithm is almost four times smaller than the average e evaluated by Poon et al [83].

Table 3.1 Relative Error in the Switching Activity from the Proposed Algorithm When Compared to the Study by Poon et al [83]

Benchmark	Relative Error [83]	Relative Error (%) [90]		
	e	e_{min}	e	e_{max}
alu4	28.94	3.58	4.78	6.88
apex2	38.05	3.92	5.78	8.17
apex4	29.98	3.66	5.02	7.67
bigkey	43.62	6.87	8.91	13.20
clma	40.53	4.68	6.06	10.30
des	38.02	2.97	4.19	6.92
diffeq	43.89	4.75	6.08	8.93
dsip	43.18	4.58	6.77	10.16
elliptic	42.49	4.05	6.00	9.30
ex1010	38.12	4.61	5.90	9.86
ex5p	33.94	3.73	4.84	8.07
frisc	42.93	6.55	8.67	12.52
misex3	23.33	3.99	5.42	8.91
pdc	38.58	3.27	5.09	7.95
s298	43.15	5.08	6.39	10.23
s38417	47.78	6.40	8.22	13.26
s38584.1	49.31	4.40	6.25	9.36
seq	36.1	3.82	5.07	7.58
spla	37.95	4.31	5.55	8.90
tseng	48.65	4.73	7.48	11.95
Average	39.427	4.498	6.12	9.51

The averages of the relative errors according to the circuit type are listed in Table 3.2. It can be noticed that the average error for datapath circuits is much less than that for nondatapath circuits. This observation agrees with the study by Schneider and Krishnamoorthy [51]. Moreover, the big gap in the signal probabilities estimation accuracy between datapath and nondatapath circuits is much less in the proposed algorithm than in the study by Poon et al [83].

In another experiment to evaluate the optimum number of cycles to be considered by the algorithm, the same experiment above was repeated for a different number of maximum cycles and the results are plotted in Fig. 3.6. It can be seen that the accuracy of the algorithm does not improve a lot after the five-cycle limit. This is mainly because the rejected cycles are those with very small probabilities, which have insignificant effect on the final probabilities. It should be noted that the error does not converge to zero with an increasing the number of cycles because of the error in power estimation in sequential circuits that is inherited from the methodology adopted for sequential power calculations.

To study the accuracy of the proposed power model in estimating the total FPGA power, the algorithm is applied to four small FPGA circuits. A brief description of the four test

Table 3.2 Relative Error in the Switching Activity from the Proposed Algorithm When Compared to the Study by Poon et al [83]

		Relative Error [90]		
	e [83]	e_{min}	e	e_{max}
Datapath circuits	34.30	3.79	5.16	8.09
Mixed circuits	44.55	5.21	7.08	10.92

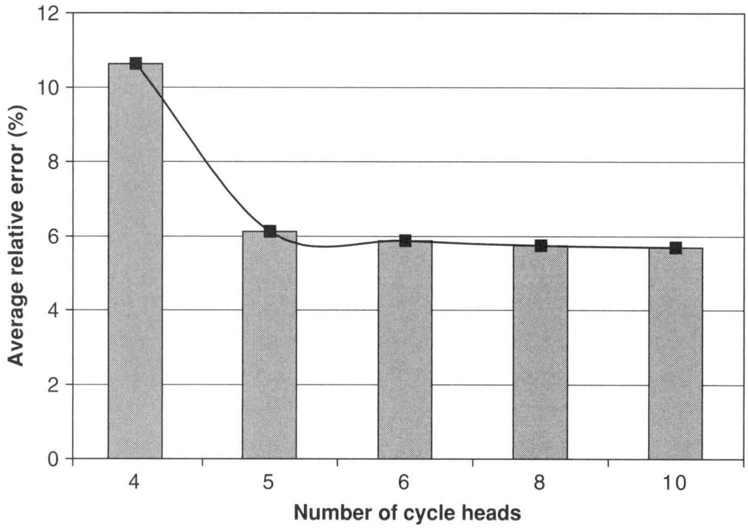

■ **FIGURE 3.6** Average relative error in estimating the signal probabilities under spatial correlation by varying the number of cycles considered.

Table 3.3 Small Benchmark Circuits

Benchmark	No. of Logic Blocks	No. of Inputs	No. of Cycles
Circuit 1	7	3	2
Circuit 2	10	5	3
Circuit 3	10	4	6
Circuit 4	14	3	8

circuits is presented in Table 3.3. Moreover, circuits 1 and 2 are datapath circuits while 3 and 4 are mixed circuits containing sequential logic and feedback loops. Moreover, the circuits are selected to feature almost no glitches to remove their effect on the results accuracy. The architecture of the target FPGA employed has a 4-input LUT and each logic cluster contains four LUTs. The different signal probabilities of all the signals in the test circuits are evaluated with and without considering spatial correlation. Afterwards, the benchmarks are designed and simulated using HSpice using a random function generator to generate the circuit inputs. The length of the stream generated by the function generator is varied from 10 to 10,000.

Figures 3.7 and 3.8 plot the average error in the signal probabilities for the case when spatial correlation is considered and when spatial independence is assumed, respectively, against the length of the input vector. The percentage error is calculated between the average signal probability calculated from the HSpice simulation and the estimated ones. In Fig. 3.7, it is noticed that the percentage error goes below 1% for an input vector of length 100. It should be noted that "circuit 2" initially has the largest error because it has a large number of inputs, five, which are not fully covered by the 10 input combinations. On the other hand, "circuit 4" has the least number of inputs, three, and hence, has the least percentage error for an input vector of length 10.

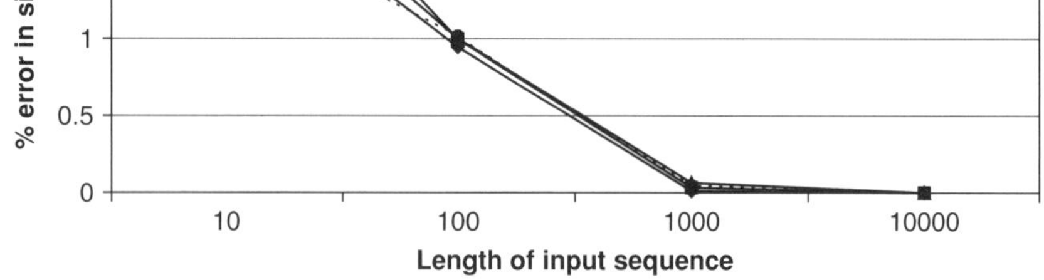

■ **FIGURE 3.7** Percentage error in estimating the signal probabilities under spatial correlation compared to HSpice versus the length of the input sequence.

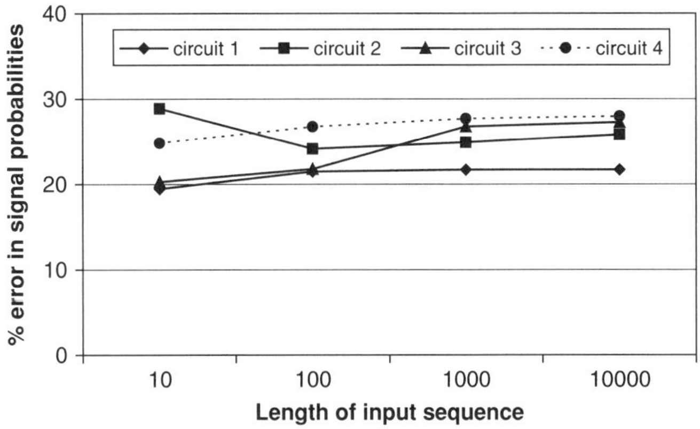

■ **FIGURE 3.8** Percentage error in estimating the signal probabilities under spatial independence compared to HSpice versus the length of the input sequence.

Figure 3.8 plots the average percentage error in estimating the signal probabilities under the spatial independence assumption. It can be noticed that the graphs for circuit 1 and circuit 4 saturate very quickly, mainly because they have the smallest number of inputs; hence, they reach their final probabilities using a small number of inputs. Circuit 2 has the maximum number of inputs that are not probably covered by a vector length of 10; thus it has the maximum percentage error for that input vector length. It should be noted that the final values of the percentage error for each circuit is due to spatial correlation. An interesting point in Fig. 3.8 is that both circuit 2 and circuit 3 have the same number of logic blocks, yet the average error in estimating the signal probabilities in circuit 3 is higher than that of circuit 2. This is because circuit 3 has more cycles than circuit 2 as well as having sequential feedback paths; hence, the error due to the spatial independence assumption is magnified.

In the next set of experiments, the same four circuits are simulated using HSpice using a CMOS 90 nm technology and their total power dissipation values are recorded. Similarly, the power dissipation in these four circuits was evaluated using the proposed power model. The total power dissipation is calculated twice, using the same equations, using transition density values computed with and without spatial correlations, and the results are plotted in Fig. 3.9. The maximum error between HSpice power calculation and that evaluated using spatial correlation is 8.8%. On the other hand, the error between the power recorded by HSpice and that calculated while assuming spatial independence is 24.2%.

In the next set of experiments, the proposed power model is used to calculate the power dissipation under spatial correlation in several FPGA benchmarks. Moreover, the same power model is used to calculate power dissipation in the same benchmarks while assuming spatial independence between the different design signals. The experiments were run on a quad Xeon processor machine running at 3.4 GHz with a total of 16 GB RAM. Table 3.4 lists the percentage difference between the power evaluated

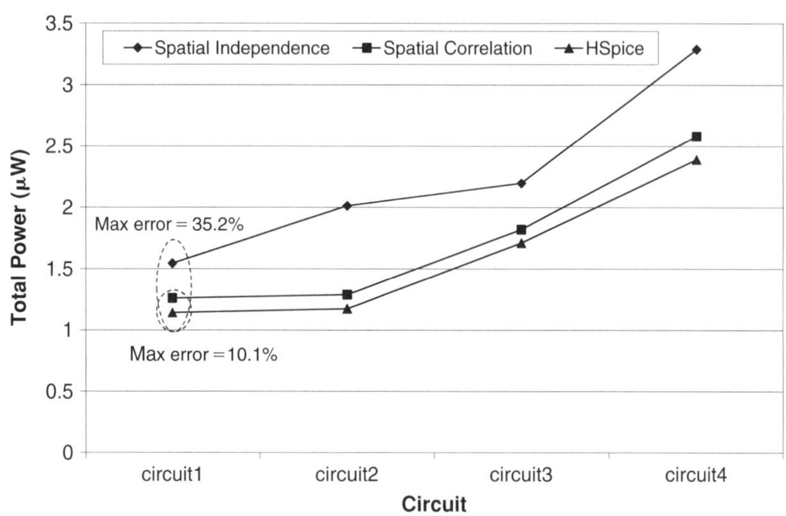

■ **FIGURE 3.9** Percentage error between power estimated with and without spatial correlation when compared to HSpice.

Table 3.4 Percentage Change in Power Estimation under Spatial Correlation Compared to Spatial Independence

Benchmark	No. of Logic Blocks	No. of Cycles	Runtime(s)	% Change in Dynamic Power	% Change in Leakage Power	% Change in Total Power
alu4	1522.00	606.00	9.00	−20.54	−26.39	−21.45
apex2	1878.00	623.00	13.00	−24.63	8.80	−20.44
apex4	1262.00	600.00	6.00	−26.54	9.83	−20.04
bigkey	1707.00	452.00	12.00	−25.82	4.84	−20.52
clma	8381.00	2343.00	614.00	−20.04	−15.74	−17.44
des	1591.00	715.00	9.00	−25.17	8.29	−19.51
diffeq	1494.00	304.00	12.00	−18.95	22.55	−9.38
dsip	1370.00	454.00	9.00	−22.97	15.19	−17.91
elliptic	3602.00	722.00	100.00	−27.29	14.24	−18.86
ex1010	4598.00	3257.00	216.00	−37.12	33.96	−24.39
ex5p	1064.00	721.00	6.00	−22.61	−19.43	−20.57
frisc	3539.00	1417.00	128.00	−26.24	21.31	−13.17
misex3	1397.00	615.00	7.00	−23.27	−26.12	−25.35
pdc	4575.00	3882.00	264.00	−18.61	−13.80	−10.15
s298	1930.00	609.00	18.00	−22.20	−11.43	−17.09
s38417	4096.00	117.00	195.00	−25.07	32.83	−16.20
s38584.1	6281.00	1396.00	281.00	−28.12	13.42	−16.90
seq	1750.00	641.00	11.00	−22.54	−33.87	−25.28
spla	3690.00	3082.00	131.00	−18.12	−30.53	−20.01
tseng	1046.00	234.00	8.00	−29.48	−16.65	−24.64

when considering spatial correlations and while assuming spatial independence. It should be noted that the runtimes reported in Table 3.4 correspond to the proposed power modeling technique with spatial correlation. Moreover, Table 3.4 lists the number of cycles found in each benchmark, which suggests that cycles are frequent in VLSI circuits; hence, spatial correlation among the different signals in a design is common. An interesting point in Table 3.4 is that the runtime is almost proportional to the number of cycles, except for two benchmarks, "clma" and "s38584.1." This is mainly because these two are the largest benchmarks and it takes a long time to process them, even without considering spatial correlation. For small benchmarks, regular probability propagation consumes considerable runtime, which covers for the increase in runtime to account for signal correlations. However, for bigger benchmarks, the runtime gets dominated by the correlation processing algorithm.

By examining the results in Table 3.4, it can be noticed that after considering spatial correlation, the dynamic power estimated for all the designs decreased because of the overestimation nature of the spatial independence assumption. On the other hand, there is no clear trend on the impact of spatial correlation among the design signals on leakage power dissipation. This is because considering spatial correlation will only change the probabilities of some of the leakage states. The state that experiences a change in its probability might be the one with the highest or lowest leakage current. Hence, there is no limitation on the change in the leakage power estimation due to spatial correlation. The average change in the total power dissipation is almost 19%.

As the CMOS process is scaled down, the contribution of leakage power dissipation to the total power dissipation is expected to increase notably until it surpasses the dynamic power by the 65 nm process [41]. To evaluate the scalability of the proposed power modeling technique with the increasing contribution of leakage power, the proposed power model is applied to all of the FPGA benchmarks in Table 3.4 using several CMOS processes (90, 65, and 45 nm). The percentage change in power estimation between the spatial correlation and independence assumptions are recorded in each case, and the average change per technology is calculated.

Figure 3.10 plots the average difference between the dynamic, leakage, and total power dissipation with and without spatial correlation. It can be deduced that the average difference in the total power dissipation estimation is almost the same for the CMOS 90, 65, and 45 nm technologies. Although, the percentage of dynamic power dissipation decreases across technologies, the impact of spatial correlation on the total power dissipation remains the same. This is because the impact of spatial correlation on leakage power dissipation stays almost the same across technologies, while the contribution of leakage power increases with the technology scaling, thus compensating for the decrease in dynamic power dissipation.

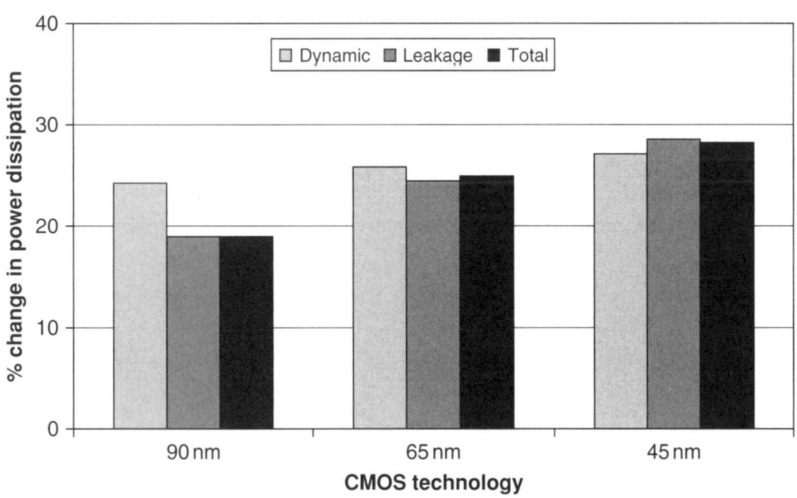

■ **FIGURE 3.10** Average percentage change in power dissipation to account for spatial correlation for different technology nodes.

This conclusion is verified by Figs. 3.11 and 3.12, which plot the similar changes broken down for leakage and dynamic power dissipation, respectively. The dependence of leakage power dissipation on the spatial correlation increases as the technology scales down, thus compensating for the decrease in dependence in dynamic power dissipation.

In another experiment, the cluster size is varied between 4BLEs, 6BLEs, and 8BLEs and the results are plotted in Fig. 3.13 for a 90 nm CMOS process. From Fig. 3.13, it can be deduced that as the cluster size increases, the impact of the spatial correlation on the power dissipation decreases. This observation can be justified by the fact that increasing the cluster size decreases the wire length of the whole circuit and the total capacitance value in (3.1). Hence, the total circuit dynamic power dissipation decreases. As a result, the dependency of dynamic power of the transition density will decrease as well. Moreover, increasing the cluster size increases the pass-transistor levels for the multiplexer in Fig. 3.4, thus resulting in a decrease in

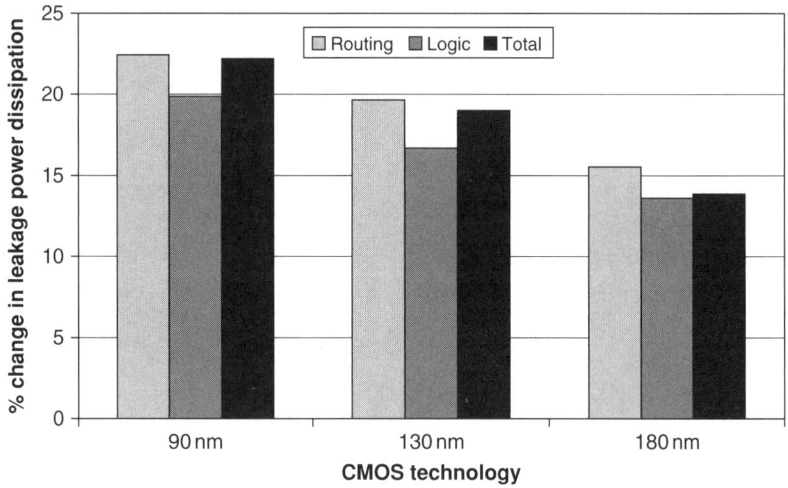

■ **FIGURE 3.11** Average percentage change in leakage power dissipation with and without spatial correlation for different technology node.

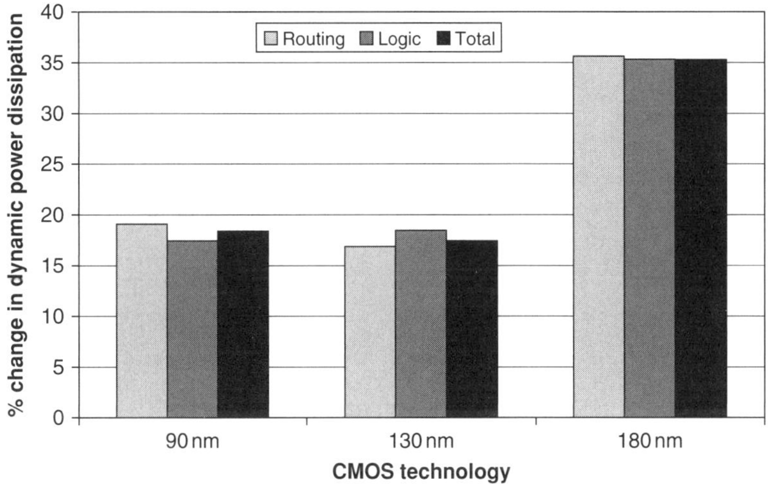

■ **FIGURE 3.12** Average percentage change in dynamic power dissipation with and without spatial correlation for different technology node.

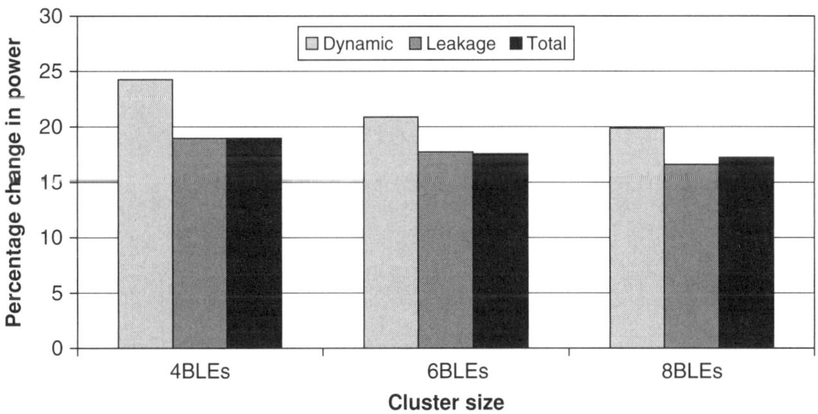

■ **FIGURE 3.13** Percentage change in power dissipation between spatial correlation and independence versus the cluster size.

the subthreshold leakage power dissipation of the multiplexers. Hence, the total leakage power dissipation of the circuit decreases significantly, resulting in a decrease in the dependency of leakage power on the signal probabilities.

Chapter 4

Dynamic Power Reduction Techniques in FPGAs

Due to historical reasons, most power reduction techniques in FPGAs developed either in academic research or in the industry focused in dynamic power reduction. The dynamic power reduction techniques can be categorized as circuit, architecture, or CAD techniques.

The first work to develop a low-energy FPGA was presented by Kurse and Rabaey [78], George et al [97], and George and Rabaey [98]. A power-optimized version of the Xilinx XC4000 FPGA was proposed, which has significant changes in the logic and routing fabrics to enable power reduction. First, larger, 5-input LUTs were used rather than 4-input LUTs to encompass more connections within LUTs instead of being routed through the FPGA interconnects. Second, an energy-efficient routing architecture was used that uses two-dimensional mesh networks, nearest-neighbor interconnects, and an inverse clustering scheme. Third, low-swing voltage interconnects were used. Last, the frequency inside the logic blocks was reduced by half by using double-edge-triggered flip-flops. In this chapter, the main techniques proposed in the literature for dynamic power reduction in FPGAs are discussed.

4.1 MULTIPLE SUPPLY VOLTAGES

The idea of dual-V_{DD} in an FPGA has been studied in various research papers [84, 86, 99–104]. The idea relies on having dual-V_{DD} supply lines in the FPGA fabric (V_{DDH} and V_{DDL}) and the CAD tool can select which one to use based on the performance and power requirements. The granularity of the selection can vary from the whole chip to a single logic module or a routing buffer. The power reduction arises from the ability to use the lower voltage V_{DD}, whenever possible, to reduce the dynamic power. Dynamic power is quadratically proportional to V_{DD}.

Commercially, FPGAs use a multiple V_{DD} line; however, there is no possible method to select which supply lines to use. The FPGA IOs are always connected to the V_{DD} line that is higher than the chip core supply line to meet certain output standards. Moreover, the SRAM cells are usually connected to a V_{DD} that is lower than the chip supply voltage for dynamic power reduction. The problems arising from the selectivity of the V_{DD} lines can be summarized as follows: (1) the need for a level converter to convert the logic output of low-V_{DD} to high-V_{DD} regions, (2) the extra hardware needed to select which supply line to use, (3) generation of the extra supply voltage, whether it will be generated internally by the chip or supplied externally from a specific pin, and (4) the need to lower the transistors, V_{TH}, to fully benefit from the lower V_{DD} without incurring too many performance penalties.

4.1.1 Predefined Dual-V_{DD} Dual-V_{TH} FPGAs

The first work that considered dual-V_{DD} for FPGAs was the study by Li et al [101]. In this work, the authors proposed a dual-V_{DD} dual-V_{TH} FPGA fabric where each logic cluster or routing resource can use either V_{DDH} or V_{DDL}. Each cluster or routing resource does not have the programmability to use any of the available supply voltages; rather the connection to the supply lines is hardwired. However, inside every logic cluster or routing resource, the configuration SRAM cells are always designed using high-V_{TH} devices for leakage power minimization.

While designing dual-V_{DD} circuits, special attention must be given to the direction of all the routed signals. When a V_{DDH} region is deriving a V_{DDL} region, the transistors in the V_{DDL} will be overdriven by a high supply voltage. As a result, these transistors will have shorter delays. However, when a V_{DDL} region is deriving a V_{DDH} region, the delays in the V_{DDH} regions will turn out to be asymmetric, since the input voltage never reaches the supply voltage V_{DDH}. Moreover, if the V_{DDL} is below, or close to, the inversion point of the V_{DDH} inverters, then the V_{DDH} inverters might suffer from signal integrity problems, and in the

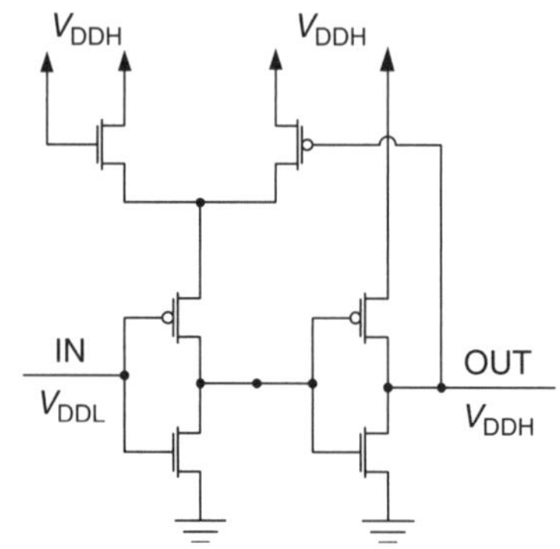

■ **FIGURE 4.1** A voltage level converter circuit [101].

worst case, might never output a logic zero. In addition, a low input voltage will also increase the short-circuit power dissipation in the V_{DDH} region inverters.

To avoid the problems associated with multiple supply voltages, level converters are needed to raise the level of V_{DDL} to that of V_{DDH}. Figure 4.1 shows the level converter used in the study by Li et al [101]. Level converters are used *only* when a V_{DDL} region is deriving a V_{DDH} region. Hence, they are only available to V_{DDL} logic resources. However, adding the level converters adds both area and delay overhead to the FPGA as well as consumes power.

In the predefined dual-V_{DD} dual-V_{TH} architecture proposed by Li et al [101], the authors proposed two FPGA fabric architectures for the dual-V_{DD} dual-V_{TH} FPGA: row-based and interleaved architectures, as shown in Fig. 4.2. The main parameter in these two architectures is the ratio between the V_{DDL} and V_{DDH} logic blocks. In all the experiments performed by Li et al [101], the ratio is set to 2:1. Hence, for every V_{DDH} logic block, the fabric has two V_{DDL} blocks. Moreover, the authors experimented with the two architectures shown in Fig. 4.2; however, they give similar results. Hence, it is recommended to use the row-based

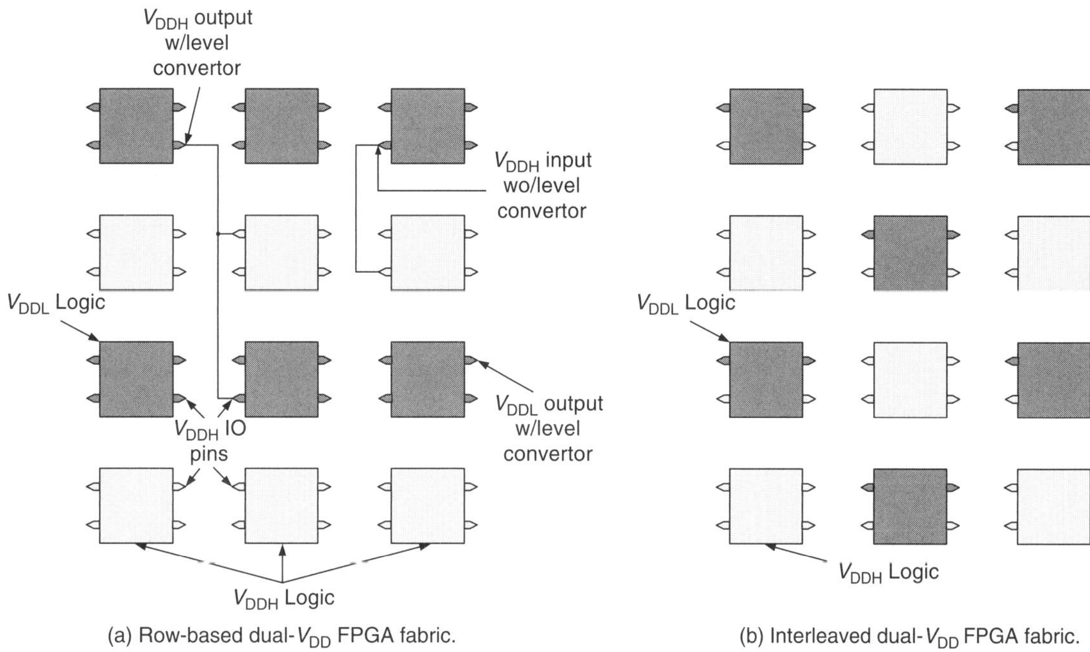

■ **FIGURE 4.2** Predefined dual-V_{DD} dual-V_{TH} FPGA fabric [101].

architecture rather than the interleaved one since it is more uniform. It should be noted that in both the architectures, all the routing is done by V_{DDH} routing resources. As a result, all the output pins of V_{DDL} logic blocks have level converters connected to them. All the input pins do not have any level converters.

One of the problems faced in dual-V_{DD} architectures is the performance reduction experienced due to using V_{DDL}. Li et al [101] investigated the impact of different V_{DD} scaling techniques on the delay of a 100 nm CMOS 4-LUT. Figure 4.3 plots the results of these experiments. Only scaling V_{DD} results in almost a three times increase in the LUT delay when V_{DD} is scaled down from 1.3 to 0.8V. Scaling down V_{TH} with V_{DD} alleviates the delay

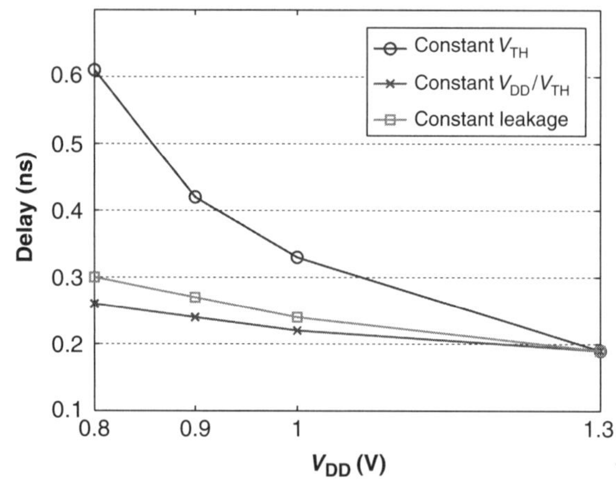

■ **FIGURE 4.3** Change in 4-input LUT delay due to reduction in V_{DD} for different V_{DD} scaling techniques [101].

increase problem, as shown in Fig. 4.3. Keeping a constant ratio between V_{DD} and V_{TH} achieves the lowest delay increase by scaling V_{DD}, only 40% delay increase when V_{DD} is scaled down by 38%. However, reducing V_{TH} by 38% results in a significant increase in the leakage power since leakage current is exponentially proportional to the decrease in V_{TH}.

Li et al [101] proposed the use of a constant leakage V_{DD} scaling. Basically, the value of V_{TH} is reduced at every value of V_{DDL} such that the leakage current of a 4-LUT operating at V_{DDL} is the same as that operating at V_{DDH}. This V_{TH} scaling technique maintains a constant leakage power of the FPGA with the reduction in V_{DD}. Using the V_{DD} scaling approach results in an increase in delay of approximately 50% with a 38% decrease in V_{DD}. The constant leakage current scaling approach provides a reasonable compromise between the constant V_{TH} and constant V_{DD}/V_{TH} supply voltage scaling techniques.

The CAD flow for the predefined dual-V_{DD} dual-V_{TH} FPGA implementation is divided into two steps. Initially, a

V_{DD} assignment phase is applied to identify which blocks will be assigned to the V_{DDL} regions. Afterwards, a supply-aware placement algorithm is applied. Finally, the resulting placed design is routed.

The V_{DD} assignment relies on a greedy algorithm that applies V_{DDL} to selected logic blocks. The algorithm starts with a V_{DDH} netlist and estimates the slacks and critical paths based on a unit delay model. A block is selected from the path with the highest slack that has the maximum power dissipation and assigned to V_{DDL}. Afterwards, the affected paths are checked to make sure that no new critical paths got created. If a new critical path gets created, or the delay on the critical path increases, that assignment is rejected; otherwise, it is accepted. The algorithm continues until it visits all the design logic blocks *once*.

Algorithm 4.1 V_{DD} assignment algorithm pseudocode [101]

Input: single-V_{DD} netlist N
Output: dual-V_{DD} netlist N'
Constraint: $crit\_path\_delay(N') - crit\_path\_delay(N) < delay\_increase\_bound$
$N' = N$
while N' has logic blocks not visited **do**
  Find path p with largest slack
  Select logic block B on p and not the critical path: $Pow_B = \max Pow(p)$
  Assign B to V_{DDL}
  if delay constraint not met **then**
    Assign B to V_{DDH}
  end if
  Mark B as visited
end while

The placement algorithm used in this study relies on the VPR CAD tool explained earlier in Chapter 1 with a slight change in the cost function. An extra term ΔCost is added to the cost and expressed as

$$\Delta\text{Cost} = \alpha \times \Delta\text{matched}(j) + \gamma \times \left(1 - \text{matched}(j)\right), \tag{4.1}$$

where matched(j) is a Boolean function that is set to 1 if block j is placed in a location with its preassigned V_{DD}, and 0 otherwise, and α and γ are weight parameters used to fine-tune the power-delay trade-off.

Li et al [101] used the V_{DD} assignment algorithm and the modified placement algorithm on several FPGA benchmarks using a V_{DDH} of 1.3 V and V_{DDL} of 0.8 V. Two architectures were tested, a single-V_{DD} dual-V_{TH} (SVDT) that only uses transistors with higher V_{TH} for the configuration SRAMs, and a dual-V_{DD} dual-V_{TH} architecture. The results of these experiments are listed in Table 4.1, where the logic resource power savings are calculated with respect to the single-V_{DD} single-V_{TH} (SVST) architecture. From Table 4.1, it can be noticed that the average power savings using the DVDT architecture is almost equal to that achieved by the SVDT. The average power savings from the SVDT architecture is 13.1%, whereas that achieved by DVDT is only 13.8%. Hence, it can be concluded that the DVDT architecture does not offer the expected power savings because the extra hardware added (level converters) consumes most of the power savings, thus rendering it impractical to use.

4.1.2 Programmable Dual-V_{DD}

In the study by Li et al [100], the dual-V_{DD} FPGA power reduction approach is extended by offering programmable V_{DD} blocks. The authors offered three different logic blocks: high-V_{DD} block (H-block) (Fig. 4.4a), low-V_{DD} block (L-Block) (Fig. 4.4b), and programmable V_{DD} block (P-block) (Fig. 4.4c). Two configuration bits are used in P-blocks to select either V_{DDH}, V_{DDL}, or power gating by

Table 4.1 Power Savings in the Logic Resources Achieved by the Dual-V_{DD} Dual-V_{TH} Architecture [101]

Circuit	SVST Power (W)	SVDT Power Savings (%)	DVDT Power Savings (%)
alu4	0.0798	8.50	14.90
apex2	0.108	9.30	7.70
apex4	0.0536	12.30	16.80
bigkey	0.148	12.30	22.10
clma	0.632	14.80	18.70
des	0.234	10.70	13.60
diffeq	0.0391	19.70	13.80
dsip	0.134	14.50	22.20
elliptic	0.14	16.30	12.00
ex1010	0.179	17.30	12.30
ex5p	0.059	11.60	16.10
frisc	0.19	19.20	18.00
misex3	0.0753	9.40	13.10
pdc	0.256	14.70	15.00
s298	0.0736	13.40	9.30
s38417	0.307	11.70	6.90
s38584	0.261	10.20	5.60
seq	0.0927	9.40	4.30
spla	0.18	12.40	22.20
tseng	0.0351	14.00	11.80

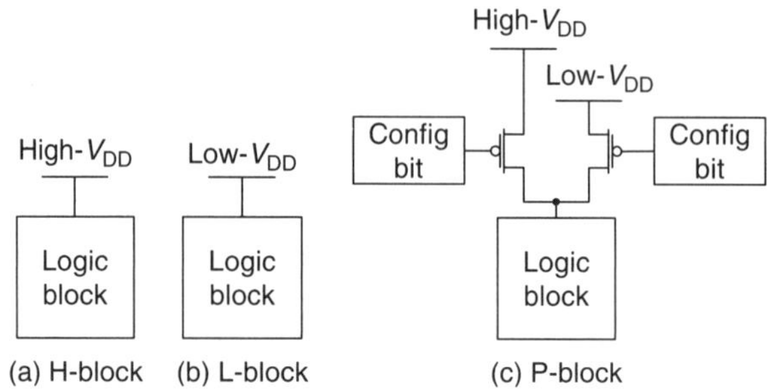

■ **FIGURE 4.4** V_{DD} programmable low-power FPGA resources [100].

turning off the two PMOS power switches if the logic block is left unused. The P-blocks in the FPGA fabric have the flexibility to either connect to V_{DDH} or V_{DDL} depending on the criticality of the net. It should be noted that in this architecture, the logic blocks still use high-V_{TH} SRAM configuration cells.

Similar to the predefined dual-V_{DD} FPGA architecture, the programmable dual-V_{DD} architecture still uses level converters, as shown in Fig. 4.1, to increase the output voltage of logic blocks operating at low-V_{DD}, either L-blocks or H-blocks. The authors also used an interleaved architecture that has for each H-block, one L-block, and three P-blocks, as shown in Fig. 4.5. It should be noted that all the routing in the fabric is still performed at V_{DDH}. The same CAD algorithm in Section 4.1.1 is used with only one slight change to the cost function in Eq. (4.1):

$$\Delta\text{Cost} = \alpha \times \Delta\text{matched}(j) + \gamma \times (1 - \text{matched}(j)) + \beta \times \Delta\text{prog}(j) + \theta \times \text{prog}(j), \tag{4.2}$$

where $\Delta\text{prog}(j)$ is a function to penalize a block moving from a matched V_{DD} region to a P-block, $\text{prog}(j)$ is a Boolean function to represent whether the current location is a P-block or not, and β and θ are parameters used to steer the trade-off between the power and delay.

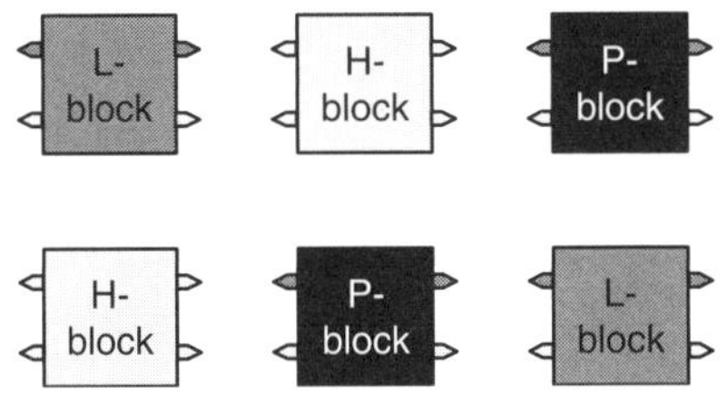

■ **FIGURE 4.5** Interleaved architecture with H-, L-, and P-blocks [100].

The power savings achieved by having programmable V_{DD} supply lines in FPGAs are listed in Table 4.2. The results are presented for having an interleaved architecture as shown in Fig. 4.5 and having another architecture with all the blocks as P-blocks. From Table 4.2, it can be concluded that although the programmable V_{DD} architecture achieves considerable power savings in the logic blocks, the savings in the total power dissipation is still not that high, only 14%. This is mainly due to the fact that most of the power dissipated in FPGAs is in the routing and IO resources. Moreover, another observation that can be made is that having configurable V_{DD} adds more power savings, thus offsetting the power dissipated in the level converters.

To increase the power savings, the authors extend the programmable dual-V_{DD} architecture to the routing resources [99]. A V_{DD}-programmable routing switch is used to select the supply voltage of the routing buffer as well as to activate the level converters, as shown in Fig. 4.6. By using programmable V_{DD} routing tracks, the total average power savings increased from 14% to approximately 50% for an FPGA with 100% P-block architectures. Table 4.3 lists the average power savings in the routing resources as well as the total power savings achieved by adopting the programmable V_{DD} FPGA architecture. One interesting point that can be deduced from Table 4.3 is that the reduction in leakage power in the routing resources is quite significant,

Table 4.2 Power Savings in the Logic Resources Achieved by the Programmable Dual-V_{DD} Architecture [100]

Circuit	DVDT (H/L/P = 1/1/3) Power Savings		DVDT (100% P)	
	Logic Power Savings (%)	Total Power Savings (%)	Logic Power Savings (%)	Total Power Savings (%)
alu4	27.06	12.66	34.20	15.83
apex4	24.33	4.18	22.18	7.58
bigkey	39.79	19.16	53.39	24.89
clma	24.78	3.07	30.07	8.82
des	46.12	10.36	56.26	19.07
diffeq	20.47	7.02	25.39	11.01
dsip	49.20	22.27	66.46	24.17
elliptic	26.49	7.89	35.10	11.62
ex5p	27.51	10.51	22.94	8.50
frisc	23.55	4.51	33.36	9.57
misex3	21.67	2.17	22.06	8.12
pdc	20.26	4.41	28.56	8.32
s298	23.36	6.21	26.32	12.87
s38417	23.01	4.45	31.27	17.45
s38584	36.34	15.47	49.88	24.99
seq	25.35	3.38	27.11	8.54
spla	28.46	15.25	32.32	14.64
tseng	27.39	9.81	41.47	21.20
Average	28.62	9.04	35.46	14.29

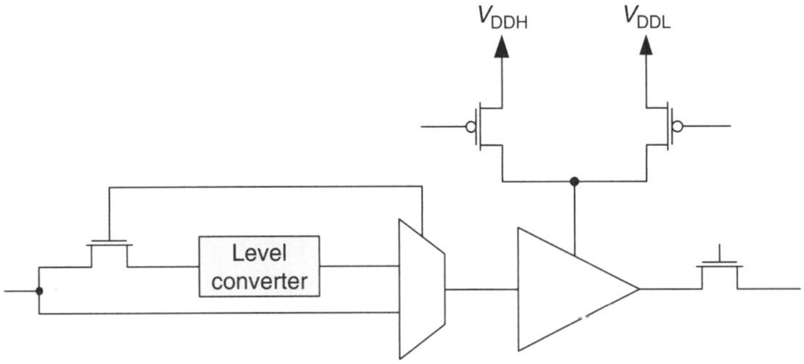

■ **FIGURE 4.6** V_{DD}-programmable routing switch [99].

approximately 80%, which arises mostly by turning off all the unused routing buffers. Most modern FPGAs suffer from a high percentage of unused routing resources.

4.1.3 **Other Dual-V_{DD} FPGA Techniques**

Several other modifications have been proposed in the literature to either increase the power savings or handle certain conditions. Lin et al [86, 102] experimented with a dual-V_{DD} FPGA architecture that contains no level converters. The FPGA fabric has duplicated routing resources, one with V_{DDH} and another with V_{DDL}. As a result, the power consumed by the level converters is saved while still benefiting from dual-V_{DD} architecture in both the logic and routing resources. However, since the locations of the V_{DDL} routing resources are predefined, the CAD tool does not have the same flexibility to achieve the maximum power savings. The average power savings from using that architecture is approximately 45%, which is slightly less than that of the programmable V_{DD} architecture. However, the main attractive part of this architecture is saving the level converters, which reduces the area of the FPGA considerably.

Another research topic for dual-V_{DD} FPGAs is performed to estimate the timing slacks on the paths accurately during the V_{DD} assignment phase of the CAD algorithm.

Table 4.3 Routing and Total Power Savings in the Programmable Dual-V_{DD} Architecture [99]

Circuit	Routing Power Savings (%)			Total Power Savings (%)
	Dynamic	Leakage	Total	
alu4	27.51	78.35	41.52	39.12
apex4	22.63	79.04	47.43	41.31
bigkey	38.85	80.61	52.51	49.12
clma	46.07	82.83	66.97	60.57
des	42.50	82.54	51.54	49.60
diffeq	39.69	77.09	64.46	52.10
dsip	40.89	84.07	57.12	57.03
elliptic	54.46	80.71	69.73	60.98
ex5p	21.24	79.87	47.15	38.45
frisc	48.12	82.65	73.06	64.42
misex3	21.83	78.40	38.07	33.43
pdc	36.15	82.02	61.78	56.37
s298	37.03	77.80	51.92	44.43
s38417	38.46	79.89	56.30	48.84
s38584	59.94	79.61	68.02	62.97
seq	26.36	79.73	43.77	38.76
spla	39.0	81.11	58.65	53.50
tseng	57.89	76.26	67.21	58.91
Average	38.81	80.14	56.51	50.55

Hu et al [103] used mixed integer and linear programming to solve that problem, whereas in the study by Lin et al [104], network flow techniques are used. Finally, Lin and He [105] used the dual-V_{DD} assignment technique under process variations to increase the yield of the resulting design.

4.2 REDUCING GLITCHES IN FPGAs

Glitching power constitutes power lost while performing unnecessary transitions. The contribution of glitches to the dynamic power dissipation can vary greatly depending on the type of the design. In arithmetic circuits, glitches contribute almost half the dynamic power dissipated. Pipelined circuits, however, have a lower contribution of glitching power since flip-flops retime the circuits to wait for slow signals. In the study by Lamoureux et al [106], the authors found out that, on average, the contribution of glitching power can reach 30% of the total dynamic power dissipated.

Several research projects have addressed glitch power reduction in ASIC circuits. Kim and Choi [107] used loop folding techniques to reduce glitching power in circuits dominated by combinational loops. The algorithm works in the synthesis phase of the CAD flow and can reduce glitching power by up to 50%.

4.2.1 Glitch Power Reduction Using Delay Insertion

Glitches occur due to unbalanced paths of inputs to combinational circuits. The most straightforward method of reducing glitches is to delay the fast input signals such that they have the same arrival times as slow input signals. Figure 4.7a shows how a delay difference between the two input signals to the XOR gate results in a glitch transition at the output. Adding an extra delay element before the fast input, which is equivalent to the difference in arrival times

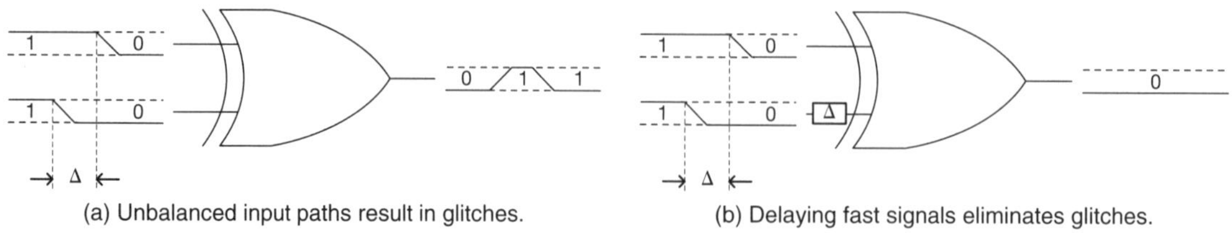

■ **FIGURE 4.7** Removing glitches by delay insertion.

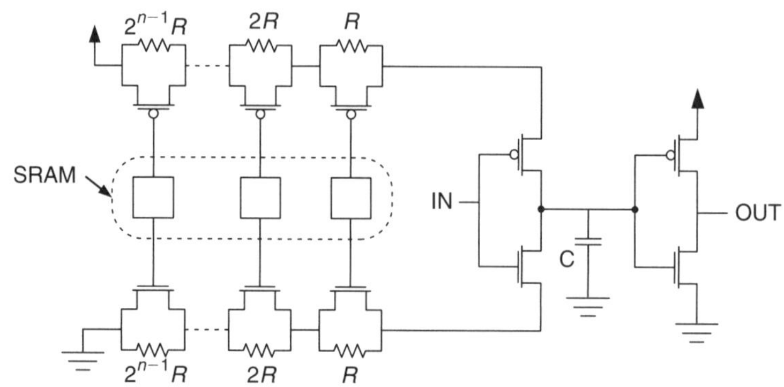

■ **FIGURE 4.8** Programmable delay elements [106].

of the two input signals, eliminates the glitch at the output, as shown in Fig. 4.7b. Lamoureux et al [106] propose adding programmable delay elements to the inputs of the logic blocks to slow down fast input signals.

Figure 4.8 illustrates the programmable delay elements used in the study by Lamoureux et al [106]. The delay element is composed of two-stage inverters, where the first one has programmable pull-up and pull-down resistors that are controlled by configuration SRAM bits. If the bypass transistor in the pull-up/pull-down network is active, the corresponding resistor is bypassed. Using the control bits, the circuit can be programmed to have any delay $\Delta \in \{k, \tau + k, 2\tau + k, \ldots, (2^n - 1)\tau + k\}$, where τ is the delay taken by the resistance R to charge/discharge the capacitor C, and k is the delay due to the bypass resistances and the inverters.

Adding such programmable delay elements to an FPGA architecture results in increasing the area as well as power dissipation of the FPGA. Hence, designers should be aware of the trade-offs involved in adding such programmable delays, or else the power savings will end up being consumed by the added delay elements. Lamoureux et al [106] investigated several architectures for the delay elements to find the one that resulted in the maximum power savings with minimum area, power, and delay overhead. The authors experimented with adding the programmable delay elements at the inputs of the logic blocks, outputs of the logic blocks, and/or inputs of the cluster. The architecture that resulted in the maximum power savings and minimum overhead is shown in Fig. 4.9, where the delay elements are placed at the inputs of the logic blocks.

Based on the architecture shown in Fig. 4.9, there are three different parameters that need to be fine-tuned to achieve the maximum power savings with minimum overhead. The first parameter is the value of the delay step that the delay element can produce (min_in). This parameter

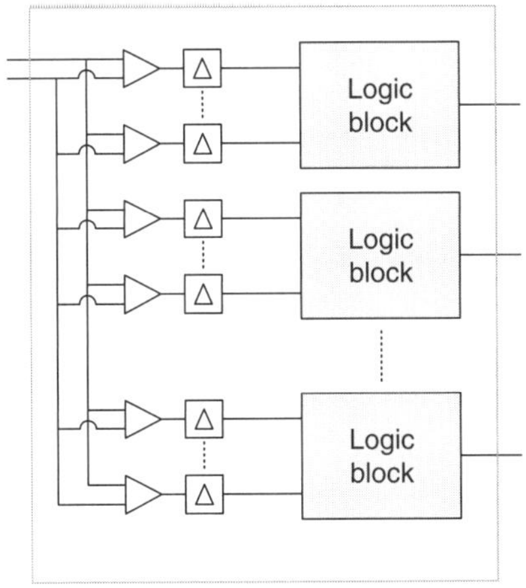

■ **FIGURE 4.9** Programmable delay FPGA architecture [106].

dictates the resolution of the delays inserted. Delay elements with a large step can balance the paths with large delay differences, but will face problems in fine-tuning small delay differences. The second parameter is the value of the maximum delay achievable by the programmable delay elements (max_in). Similarly, the maximum achievable delay affects the delay matching capabilities of the delay element. The third parameter is the number of delay elements used per LUT (num_in).

Lamoureux et al [106] proposed a delay assignment CAD flow to determine the configuration of each delay element in the architecture. For each fanin f to every logic block n in the design, the needed_delay is calculated according to

$$\text{needed\_delay} = \text{arrival\_time}(n) - \text{arrival\_time}(f) - \text{fanin\_delay}(n,f). \tag{4.3}$$

Afterwards, the added_delay for each fanin to every logic block is calculated using the delay assignment Algorithm 4.2.

Algorithm 4.2 Delay assignment algorithm for delay insertion glitch reduction [106]

for each LUT $n \in$ circuit **do**
  $count = 0$
  for each fanin $f \in n$ **do**
    if $((needed\_delay(n,f) > min\_in)$ && $(needed\_delay(n,f) \leq max\_in)$ && $(count < num\_in))$ **then**
      $added\_delay(n,f) = min\_in \times \lfloor needed\_delay(n,f)/min\_in \rfloor$
      $needed\_delay(n,f) = needed\_delay(n,f) - added\_delay(n,f)$
      $count++$
    end if
  end for
end for

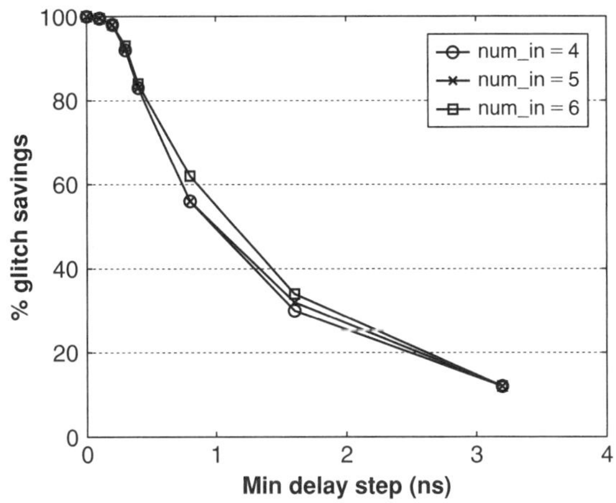

■ **FIGURE 4.10** The impact of the delay step on glitch power savings [106].

To find out the optimum values of the three design parameters (min_in, max_in, num_in), several experiments were performed in the study by Lamoureux et al [106]. In the first experiment performed, max_in is set to ∞, num_in is set to the number of inputs to the LUT, and min_in is varied, as shown in Fig. 4.10. From Fig. 4.10, it can be noticed that having a very small value for min_in increases the ability to match any delay differences, especially with max_in of ∞, thus increasing the glitching power savings. However, this comes at the expense of significant area and power overhead. Moreover, a min_in of 0.25 ns is sufficient to get rid of most of the glitching power, as shown in Fig. 4.10. It should be noted that glitches that propagate through the fabric are usually wider than the delay of the logic block. Hence, there is no point in min_in smaller than the delay of one logic block.

In the next experiment, min_in is set to 0.25 ns, num_in is set to the number of inputs to the LUT, and max_in is varied, as shown in Fig. 4.11. It can be observed that increasing max_in increases the glitching power savings; however, the savings reach a plateau beyond 12 ns.

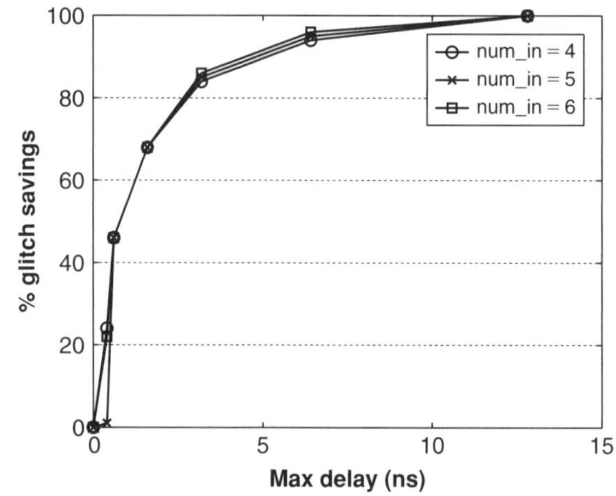

■ **FIGURE 4.11** The impact of max delay on glitch power savings [106].

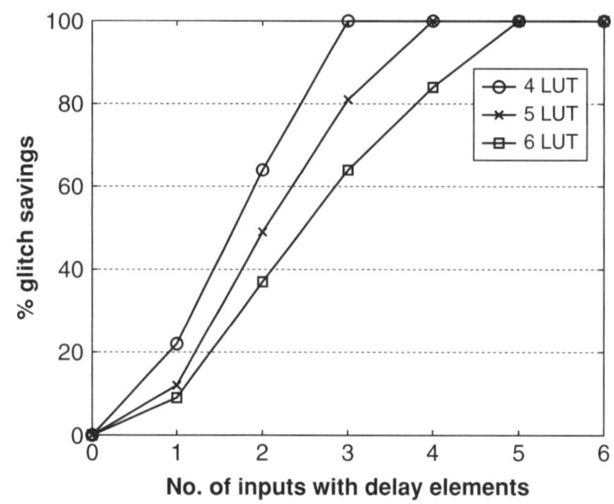

■ **FIGURE 4.12** The impact of number of delay elements per LUT on glitch power savings [106].

In the last experiment, min_in is set to 0.25 ns, max_in is set to 8.0 ns, and num_in is varied for different LUT sizes, as shown in Fig. 4.12. From Fig. 4.12, it can be concluded that setting num_in to one less than the size of the LUT is enough to achieve considerable power savings without incurring too much area overhead. Table 4.4 lists

Table 4.4 Power, Area, and Delay Overhead Due to the Programmable Delay Elements [106]

LUT Size	Average Power Overhead (%)	Average Area Overhead (%)	Average Delay Overhead (%)
4	0.89	5.3	0.21
5	0.94	5.0	0.13
6	0.98	4.4	0.14

the average power, area, and delay overhead due to the programmable delay elements using the aforementioned values for the three design parameters. From Table 4.4, it can be concluded that the delay insertion glitch power reduction has the least overhead for large LUT sizes. Finally, Table 4.5 lists the overall power savings due to delay insertion in several FPGA benchmarks. The average savings in power dissipation is approximately 18%.

One of the main issues using this approach, other than the increase in area, is the predictability of the delay resulting from the delay element with the variations in the operating temperature, process parameters, and supply noise. Several research projects targeted designing more robust and accurate programmable delay elements [108–110].

4.2.2 Multiphase Flip-Flop Insertion for Glitch Power Reduction in FPGAs

Long chains of cascaded logic blocks usually result in glitches due to path delay differences between the input signals. Inserting flip-flops to cut these long chains terminates the propagation of glitches to consecutive logic cells. Lim et al [111] proposed inserting flip-flops in long logic chains using an intelligent clock assignment technique for the added flip-flops.

Table 4.5 Overall Power Savings Due to the Programmable Delay Elements [106]

Circuit	Power Savings (%)
C135	25.4
C1908	18.1
C2670	11.6
C3540	27.5
C432	13.0
C499	31.8
C5315	18.2
C6288	52.1
C7552	22.6
C880	7.2
alu4	2.5
apex2	3.6
apex4	9.5
des	15.1
ex1010	16.8
ex5p	23.8
misex3	7.6
pdc	11.1
seq	5.3
spla	20.3
Average	18.2

The methodology proposed in the study by Lim et al [111] uses the unused flip-flops inside the basic logic cell to cut long combinational chains. Since this methodology uses the unused flip-flops, the area of the design does not increase. To avoid any change in the design, the flip-flops have to be clocked using phase-shifted clocks instead of the original design clock.

An example of glitch filtering using the unused flip-flops is shown in Fig. 4.13, where the gray wires represent disabled paths. FF1 in Fig. 4.13a blocks all the glitches generated by L1 or any other logic block in its fanin. In Fig. 4.13b, FF1 is disabled; hence, glitches generated by L1 or any of its

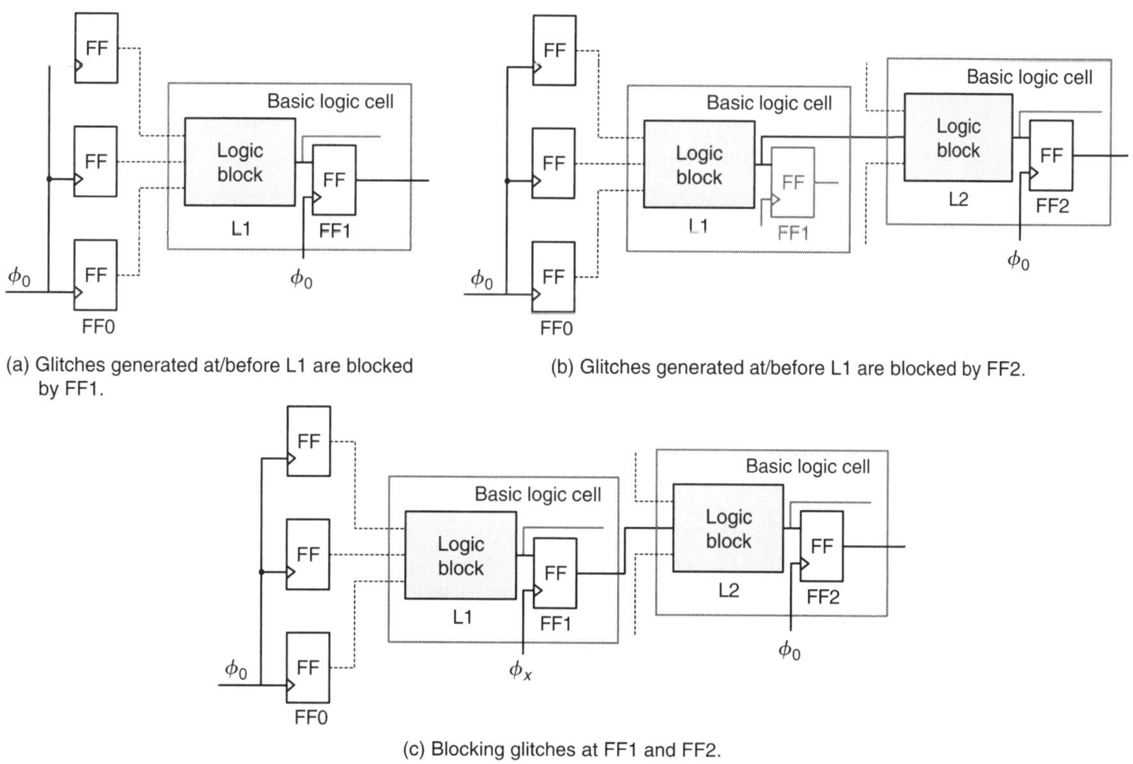

■ **FIGURE 4.13** Glitch blocking by flip-flop insertion [111].

fanin blocks propagate to the output of L2, where they are blocked by FF2. As a result, the configuration in Fig. 4.13b consumes more power than the one in Fig. 4.13a. In Fig. 4.13a, the unused FF1 is used for blocking the glitches at the output of L1, thus preventing them from propagating to L2. However, if FF1 runs at the same clock ϕ_0 as FF2, the result of this path will be available after two clock cycles. If FF1 runs at a clock with a different phase but same frequency as ϕ_0, the result of this path can be evaluated at one clock cycle. Determining the phase difference between ϕ_0 and ϕ_x depends on the slack and delays of the path.

The glitch blocking technique comes at the expense of power overhead in other parts of the design. The added flip-flop adds capacitance to the signal path, thus consuming dynamic power. Moreover, the phase-shifted clock generated also consumes dynamic power through the clock network. Hence, the power savings achieved from blocking the glitches are slightly reduced by the power overhead. However, the most interesting aspect of this technique is that it does not require any architecture modifications; hence, the designer can only use when reducing the glitches is pivotal to reducing the total power dissipated in the design.

There are some limitations to applying this methodology for glitch blocking. First, the design needs to have some unused flip-flops in the used logic resources. Moreover, these unused flip-flops should not lie on the critical path and have sufficient slack to use for the phase-shifted clock. Second, the design should have available clock routing resources for the added clock. Moreover, the phase shift needed should be within the resolution of the FPGA phase-locked loop (PLL) or delay-locked loop (DLL). In addition, the number of clocks used should be as small as possible to minimize the power overhead. As a result, an optimization algorithm is needed to solve this problem.

Lim et al [111] proposed the flip-flop and phase-shifted clock assignment (FPA) algorithm to identify the flip-flops

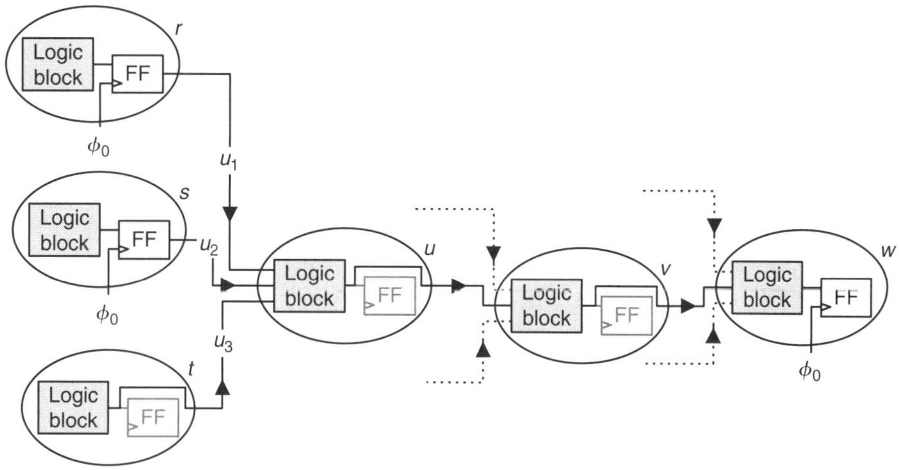

■ **FIGURE 4.14** A DAG representation of an FPGA circuit [111].

that will be activated in the design and the number of clock phases needed. The FPA algorithm is performed on the postlayout netlist. The algorithm starts by converting the design into a directed acyclic graph (DAG), *G*, where every LUT in the design is mapped to a vertex and nets are mapped to edges, as shown in Fig. 4.14. Each LUT and its following flip-flop are converted to a single vertex. Afterwards, the FPA algorithm calculates the weight *w* of each vertex in *G*, which is a representation of the amount of power savings achievable by activating the unused flip-flop at each vertex.

To properly calculate the value of *w*, both the total capacitance seen at each vertex and the glitching power transition density should be calculated, since power is proportional to the capacitance and transition density. As an example, consider the DAG in Fig. 4.14. All the glitches that are generated in the circuit before *r* and *s* are blocked by the flip-flops clocked by the in-phase clock ϕ_0 and are not propagated into *u*. However, assume that *t* does not generate any glitches and due to the input arrival time difference at *u*, glitches are generated at its output. Hence, activating the unused flip-flop in u results in saving glitching power that will charge/discharge the output capacitance of u ($C_{\text{eff}}(u)$). Furthermore, the glitch generated in u will propagate into its fanouts, i.e., v, resulting in more glitching power. This trend will continue until the path is terminated by an in-phase flip-flop, as in w. Consequently, in calculating the power saved by activating the unused capacitance at u, the capacitance of all its fanouts should be accounted for in a depth-first approach until a flip-flop is reached. Hence, the total capacitance at u is given in the study by Lim et al [111]:

$$C_{\text{total}}(u) = C_{\text{eff}}(u) + \sum_{v \in \text{fanout}(u)} C_{\text{total}}(v) \times P\left(\frac{\partial y}{\partial \overrightarrow{uv}}\right), \quad (4.4)$$

where $\overrightarrow{uv}$ is the edge that goes from u to v, y is the output of v, and $P\left(\frac{\partial y}{\partial \overrightarrow{uv}}\right)$ is the Boolean difference between y and $\overrightarrow{uv}$ and calculated using Eq. (3.10). The Boolean difference is used in Eq. (4.4) to account for only the amount of glitching generated at the output of u that will propagate to the output of v based on the logic function implemented by v.

In Chapter 3, the transition density at any vertex is defined using Eq. (3.9). Lim et al [111] propose another formulation for the transition density at y with n inputs $x_1, \ldots, x_n$ to account for the circuit delays

$$D'(y) = \min\left[\sum_{i=1}^{n} P\left(\frac{\partial y}{\partial x_i}\right) D'(x_i), \lceil \tau/\mu \rceil\right], \quad (4.5)$$

where τ is the circuit clock period and μ is the delay due to one LUT. The minimum of both quantities is used in Eq. (4.5) to reject all transitions that are narrower than the LUT delay. It should be noted that the formulation presented in Eq. (4.5) is a crude approximation of the actual transition density due to LUT delays. Afterwards, the authors [111] estimate the transition density due to glitches as

$$G_{\text{trans}}(y) = D'(y) - D(y). \tag{4.6}$$

Using Eqs. (4.4) and (4.6), weight at u is calculated as [111]

$$w(v) = G_{\text{trans}}(u) \times C_{\text{total}}(u). \tag{4.7}$$

The next step in the FPA algorithm is to calculate the slack available for each vertex in the DAG to get an estimate of the freedom available for adding the new clock domains. The slack in the study by Lim et al [111] for vertex v is calculated based on the assumption that all the fanin signals of v arrive as early as possible and all the fanouts of v arrive as late as possible. Consequently, if the cumulative delay after v is τ_f and the cumulative delay before v is τ_b, then the allowable phase range available for v is given by

$$\xi = \left[\frac{2\pi\tau_f}{\tau}, \frac{2\pi\tau_b}{\tau}\right] \iff \tau > \tau_f + \tau_b, \tag{4.8}$$

where τ is the clock period. Figure 4.15 shows an example of the slack distribution of a DAG with the weights on every edge. The different clock phases available are denoted by a, b, c, d, e, and f, with the difference between their phases given by p, which corresponds to the FPGA PLL/DLL resolution. The weight, or power savings, of each clock phase is

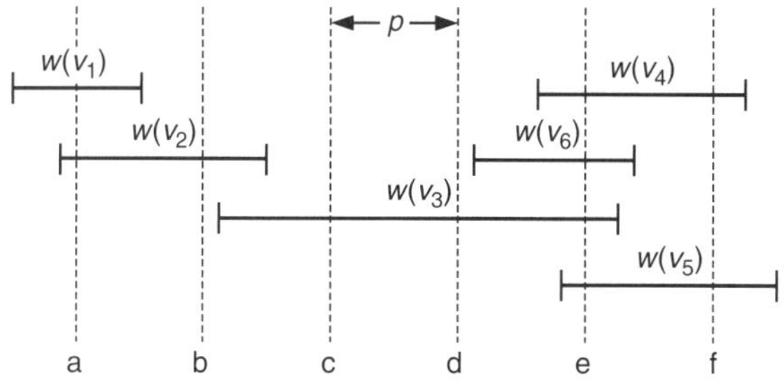

FIGURE 4.15 Weighted slack intervals [111].

calculated by summing up the weight on each clock phase. As an example, the phase weight, pw, of phase e is given by

$$pw(e) = w(v_3) + w(v_4) + w(v_5) + w(v_6). \tag{4.9}$$

The clock phase that maximizes the power savings is the one that has the maximum pw (mpw). However, getting the phase that maximizes the power savings is not as simple as getting the one with the maximum pw because some of the vertices might have a direct connection. Hence, their ξ will have a strong correlation between them. It was proven by Lim et al [111] that finding the maximum pw is an Non-deterministic Polynomial (NP) problem because of the correlation between the different vertices in the DAG.

To identify the phase that will result in the maximum power savings, Lim et al [111] used a greedy algorithm to solve that problem. For every phase p, the algorithm sorts all the set of weights V_p descendingly. All the vertices in V_p that are connected to the vertex with the maximum weight are then deleted from V_p to remove the correlation between the vertices that belong to a certain phase. The same procedures are repeated for the next vertex/weight in V_p until all the V_p is processed. The pw is then calculated using the weights remaining in V_p. The algorithm then considers the phase with the maximum weight and calculates the power overhead resulting from activating the flip-flop(s) associated with it. If the power savings are enough to offset the power overhead, the flip-flop(s) are activated; otherwise, not. The pseudocode of the FPA algorithm is listed in Algorithm 4.3.

The FPA algorithm was applied on several benchmarks to calculate the amount of power savings that can be achieved from flip-flop insertion in FPGAs. Table 4.6 lists some of the results reported in the study by Lim et al [111] for the FPA algorithm. The second column in Table 4.6 lists the number of clusters for each design, while the third column lists the number of phases that are evaluated by the FPA algorithm for each benchmark. It can be observed that most of the benchmarks needed only

Algorithm 4.3 FPA algorithm pseudocode [111]

```
Input: FPGA design D, c = (c_1, ..., c_{n_c}), p = (p_1, ..., p_{r_p})
Output: power minimized FPGA design D_new
get graph G from design D
for each new clock c_i do
  for each v ∈ G do
    get weight w and interval ξ
  end for
  for each phase p_j do
    get maximum phase weight mpw(p_j)
  end for
  select p_max with maximum mpw(p_j)
  if mpw(p_j) ≤ overhead then
    break
  end if
  assign p_max to c_i
  modify G with c_i
end for
get D_new from G
```

one clock phase to achieve the maximum power savings, except for two benchmarks. The other observation that can be made based on the clock phases selected by the FPA algorithm is that most of them are multiples of 45° with the majority of them close to 135° and 180°.

The fifth column in Table 4.6 lists the total number of candidate flip-flops, or unused flip-flops, available in the circuit and the number of unused flip-flops that ended up being activated by the FPA algorithm. The interesting observation that can be made from this column is that the ratio of selected to candidate flip-flops is not always proportional to the glitch savings achieved; the same glitch savings was achieved by activating only 9% of the candidate flip-flops in C499, while approximately 30% were needed in the Multiplier14 circuit. This is mainly

Table 4.6 Power Savings for Flip-Flop Insertion Using the FPA Algorithm [111]

Circuit	No. of Clusters	No. of Phases	Phases Selected	No. of Nodes (Selected/ Candidate)	Glitch Reduction (%)	Total Power Reduction (%)
Adder14	133	1	261.5°	217/228	95.47	8.67
Multiplier14	1080	1	187°	606/2000	56.78	20.63
C432	907	1	184°	444/1786	60.55	4.31
C499	331	1	163°	59/650	53.38	13.41
C6288	1045	6	163°, 117°, 218°, 138°, 263°, 80°	264/1990	63.72	31.74
Tap16-D	1175	3	150°, 165°, 325°	321/1648	57.33	12.74

because the amount of glitch power savings does not depend on the number of flip-flops activated, but it rather depends on the weight of the vertex activated. On average, the FPA algorithm achieves 64% glitch power savings and 15% total power savings, thus suggesting that the glitch power contribution in the circuits tested to be around 25%.

Figure 4.16 plots the dynamic energy of several designs tested with increasing the number of phases in the design. It can be observed from Fig. 4.16 that increasing the number of phases beyond two usually results in either increasing the power dissipation in most of the circuits, except for C6288. Moreover, the C6288 circuit witnesses a reduction in the power dissipation at three clock phases, but the reduction in power seems to flatten after that. Most of the power savings achieved from the C6288 circuit actually arise from using two clock phases. Hence, it can be concluded that having two clock phases is enough to achieve the maximum power savings from the flip-flop insertion scheme. Moreover, setting the two clock phases to 135° and 180° is enough to achieve the maximum power savings.

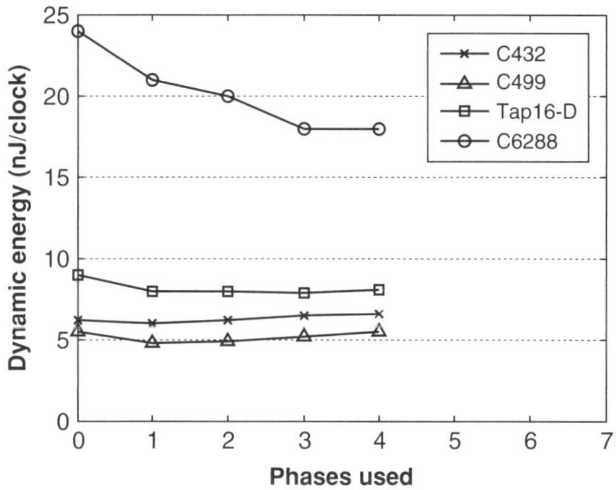

■ **FIGURE 4.16** Energy savings with increasing the number of clock phases used [111].

4.2.3 Negative Edge Flip-Flop Insertion for Glitch Power Reduction in FPGAs

As discussed in Section 4.2.2, adding flip-flops controlled by out-of-phase clocks adds considerable power overhead. Most of the power overhead is due to the power dissipated in the clock resources that belong to the newly added clocks. This is mainly why using a large number of clock phases does not end up in achieving more power savings. Czajkowski and Brown [112] proposed using a single phase flip-flop insertion scheme. The authors used negative edge-triggered flip-flops, i.e., a phase shift of 180°.

Figure 4.17 illustrates an example of a negative edge-triggered flip-flop. In the circuit in Fig. 4.17, block C will experience glitches due to the difference in delays of its input signals and that glitch will propagate into D. Inserting a negative edge-triggered flip-flop at the output of C will block all the glitches that occur during the first half of the clock cycle. As it was observed from the results in the study by Lim et al [111], having a phase shift of 180° was sufficient to block all the glitches in the circuit. In the next

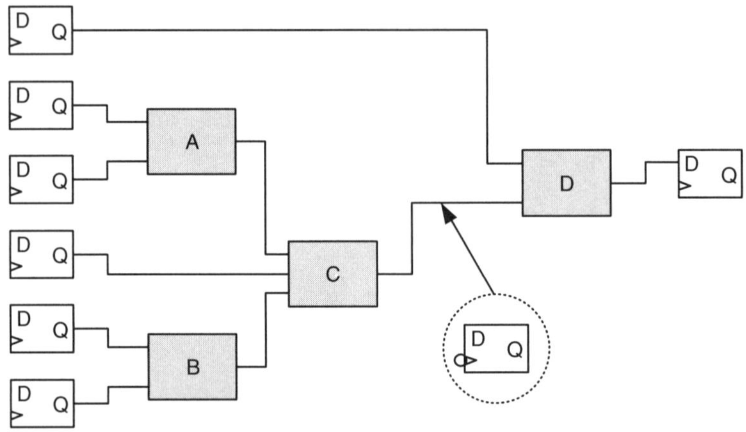

■ **FIGURE 4.17** An example of negative edge flip-flop insertion [112].

half of the cycle, the added flip-flop will latch the data on its inputs and keep it for another cycle.

In inserting negative edge-triggered flip-flops, the authors used a very simple CAD techniques that place a negative edge flip-flop at every node in the circuit that might experience glitches such that the power savings exceed the power cost of adding the flip-flop. Table 4.7 lists the impact of the negative flip-flop insertion on the area, delay, and power dissipation. As can be observed from Table 4.7, the impact of the negative flip-flop insertion on the area and delay of the final design is minimal.

Moreover, it can be noticed that the change in power dissipation reported in Table 4.7 is listed as positive for some of the circuits tested, or close to zero. This observation contradicts the fact that the CAD flow used in the study by Czajkowski and Brown [112] ensures that the power overhead of the flip-flop should be less than the power savings achieved by adding the flip-flop. As a result, it can be concluded that the power model used for glitches in the study by Czajkowski and Brown [112] is not very accurate in estimating the savings due to glitches. This observation also justifies the small average power savings, only 7%, reported by the authors.

Table 4.7 Power Savings Using Negative Edge Flip-Flop Insertion [112]

Circuit	Change in Area (%)	Change in Delay (%)	Change in Power (%)
barrel64	1.46	8.74	−17.5
mux64_16bit	0	0	0
fip_cordic_rca	0	3.82	−8.76
oc_des_perf_opt	0.96	2.64	−24.75
oc_video_comp_sys_huffman_enc	0.93	0	0.33
cf_fir_24_8_8	0	5.87	0.84
aes128_fast	2.23	4.84	−0.99
rsacypher	0	2.85	−4.95
Average	0.7	3.6	−7.00

4.2.4 Behavioral Synthesis with Flip-Flop Insertion for Glitch Power Reduction in FPGAs

As it was shown in Sections 4.2.2 and 4.2.3, flip-flop insertion can significantly reduce the power dissipation due to glitches. However, the results presented in Sections 4.2.2 and 4.2.3 showed small total power savings. The main reason for the meager power savings is that the flip-flop insertion was applied late in the CAD algorithm at the placed and routed design. Hsieh et al [113] aimed to apply the flip-flop insertion scheme high up in the CAD flow at the synthesis level. They insert an in-phase register at the output of functional operations that will generate glitches [113]. However, using the in-phase clock might result in data hazards in the circuit output; hence, extreme caution must be taken in identifying the location of the inserted register.

The authors divide the circuit into functional units [113] instead of single LUTs as presented in Section 4.2.4. The registers are inserted at the boundaries of functional units.

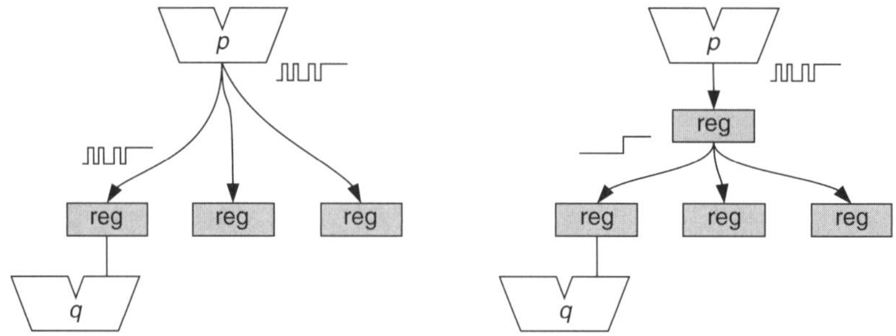

(a) Glitches generated by *p* propagate to *q*. (b) Glitches generated by *p* are blocked by adding a register at the boundary of *p*.

FIGURE 4.18 Glitch blocking by register insertion on the boundaries of functional units [113].

This scheme will result in the maximum power savings when the boundary signals of the functional units feed more than one functional unit, as shown in Fig. 4.18. Adding a register to the boundary output of p blocks the glitches from propagating in the long routing resources that connect p to all the next functional units, including q, as shown in Fig. 4.18b. However, adding the register to the boundary of p may affect the correctness of the design functionality.

Consider the circuit and its scheduled data-flow graph (DFG) representation in Fig. 4.19a. A DFG is basically a DAG, where every vertex represents an operation and an edge represents a dataflow from one vertex to another. Scheduling the DFG is identifying which operations are executed at the same time step, or control step (c-step). In Fig. 4.19a, functional units p and q are bound to operations u and v, respectively. Activating the unused register at the output of p, as shown in Fig. 4.19a, makes the result of u available to v after two c-steps instead of one as the DFG states, resulting in a write-after-read data hazard. Solving the data hazard problem at the input of q can be achieved by forwarding the result of p directly into q without going through the register before q, as shown in Fig. 4.19b. Hence, the result of u will be available to v after

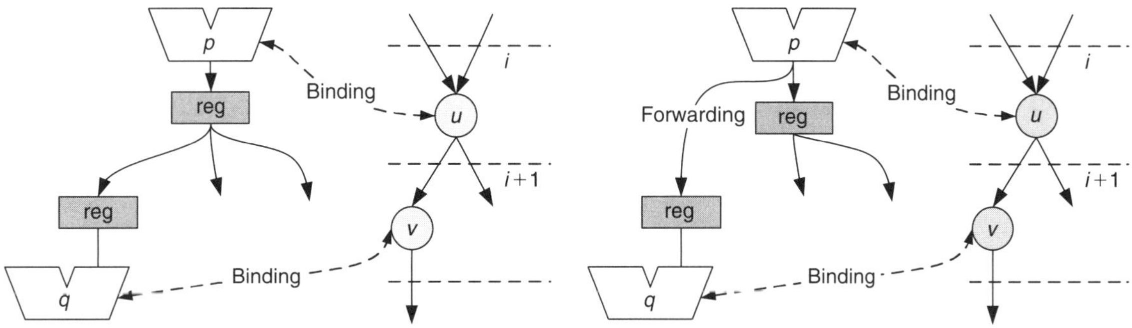

(a) Inserting a register at the boundary of p results in data hazards at the input of q.

(b) Forwarding is used to reslove the data hazards.

FIGURE 4.19 Data hazards and forwarding [113].

only one c-step, as required. The added register should keep the data of u until v is done reading it. If v needs the data for more than one clock cycle, it might create data hazard problems at v.

Hsieh et al [113] presented a formulation for the data hazard problem and the conditions needed for data hazard to occur. Data hazards will occur on a dataflow u to v if and only if (1) v is a nonpipelined multicycle operation that is executed at k consecutive c-steps, labeled $\{i, i+1, \ldots, i+k-1\}$, (2) operations u and v are executed at consecutive c-steps, and (3) there exists an operation w such that u and w are bound at the same function unit and w produces results between c-step i and $i+k-2$. Hence, registers should not be used whenever a data hazard is expected to occur. Figure 4.20 shows an example of a data hazard in flip-flop insertion.

A resource binding algorithm that tries to avoid the conditions that result in data hazard was proposed [113], as explained above. Traditionally, binding algorithms target power minimization by reducing the switching activities of the resources in the circuit [114, 115] without considering the flip-flop insertion for glitch reduction. The main idea of the binding algorithm proposed by Hsieh et al [113] is

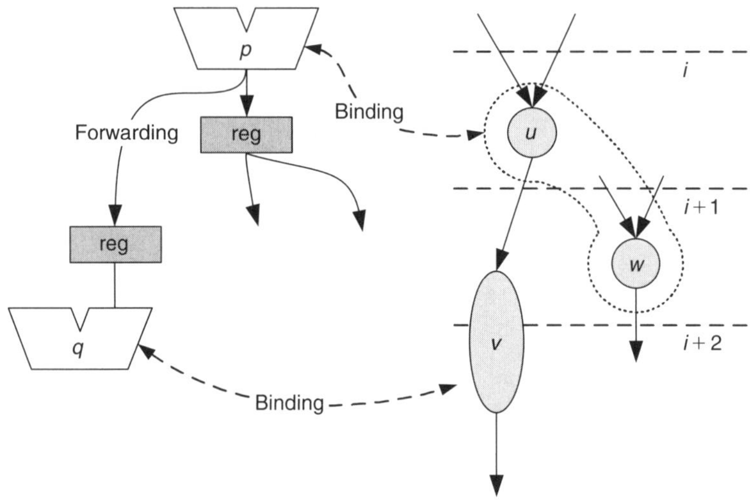

■ **FIGURE 4.20** An example of a DFG that creates data hazard [113].

to bind any two operations that depend on each other into two nonconsecutive c-steps. As an example, in Fig. 4.20, the flip-flop insertion binding algorithm will try to move operation v such that it is not executed in nonconsecutive c-steps to u.

To move an operation that takes more than one c-step to later c-steps, the slack of each dataflow involved is used. The slack of a dataflow edge (u, v) is defined as the number of c-steps between operations u and v. Positive slacks indicate that the two operations can be executed at nonconsecutive c-steps, whereas a zero slack indicates consecutive operations. Hence, to fully use from the flip-flop insertion at the synthesis stage, whenever a data hazard is expected to occur, the slacks of the consecutive operations can be used to move the conflicting operations into nonconsecutive c-steps. The authors [113] use a network flow model to solve that problem optimally.

Table 4.8 lists the power savings achieved from the behavioral synthesis flip-flop insertion technique. The results

Table 4.8 Power Savings Behavioral Synthesis for Flip-Flop Insertion [113]

Circuit	ADD/SUB	MUL	Final Stage Register Insertion			Register Insertion Synthesis		
			Change in FFs (%)	Change in LUTs (%)	Change in Power (%)	Change in FFs (%)	Change in LUTs (%)	Change in Power (%)
ARAI	6	1	14.20	−0.60	−12.33	16.57	10.32	−18.49
DIF	6	2	8.12	0.10	−10.92	16.24	4.26	−27.59
DIT	7	3	8.58	−3.38	1.51	15.45	6.75	−29.65
LEE	6	4	6.23	0.53	−9.20	14.01	2.66	−33.91
MCM	13	6	18.40	8.44	0.83	17.43	6.71	−8.71
WANG	5	4	9.57	2.71	−29.41	17.22	6.71	−34.80
CHEM	33	33	16.08	1.04	−23.09	12.61	−12.09	−22.67
DIR	11	12	20.32	3.11	−29.09	21.25	2.92	−41.82
HONDA	9	10	22.29	−1.53	−21.52	22.29	−0.23	−25.56
PR	5	3	11.68	1.51	−26.23	23.36	−1.05	−37.23
Average			13.55	1.19	−15.95	17.64	2.7	−28.00

are listed for two different schemes. In the first scheme, flip-flops are only inserted at the final stage of the synthesis phase when data hazards are not expected to occur. Slacks are not used in that scheme to add further flip-flops. In the second scheme, slacks are used to move operations between c-steps to add more flip-flops without causing data hazards. The results in Table 4.8 show an average power savings of 16% by inserting flip-flops when data hazards will not occur. This average power savings is very close to the previously discussed glitch power reduction techniques. Using slacks during synthesis to add more flip-flops to the circuit resulted in approximately 28% power reduction. As a result, it can be concluded that addressing the flip-flop insertion problem at a higher design level can achieve more power savings.

However, behavioral synthesis for flip-flop insertion results in an area increase of approximately 4%. The main reason for this increase in area is that at the synthesis stage, adding an extra flip-flop to the circuit does not guarantee that it will go to an unused flip-flop location in the FPGA fabric. The packing and placement stages, which come after synthesis, are responsible for that. The synthesis phase does not have any control of the location of the inserted flip-flop.

4.3 CAD TECHNIQUES FOR REDUCING DYNAMIC POWER IN FPGAs

4.3.1 Power Reduction Techniques during Technology Mapping

During the technology mapping phase in FPGAs, the circuit is transformed into a group of LUTs and flip-flops. Traditionally, FPGA mapping techniques aim to optimize the circuit area [116–118] and/or the resulting design depth [119–121]. Several works in the literature addressed power minimization during technology mapping, in addition to minimizing area and/or depth [122, 123].

4.3.1.1 ***Power-Aware Technology Mapping***

In the study by Anderson and Najm [122], a mapping algorithm is proposed that tries to minimize the number of highly active nets that are passed to the routing resources. Moreover, the authors offer a power-aware logic replication technique.

During mapping, the circuit is converted into a DAG where each node represents a single-output logic function and edges between nodes are the input/output dependencies between the different logic functions. Before explaining the power-aware technology mapping technique developed by Anderson and Najm [122], the terminology used in the study will be introduced. For a subgraph H, input(H) is a set of all the fanins of nodes in H. Similarly, output(H) contains all the nodes that do not belong to H and are fanouts of the nodes in II. A subgraph is said to be rooted at a node z if the subgraph contains z and all of its predecessor nodes, i.e., all the fanin nodes of z. A K-feasible cone at z (F_z) is defined as a subgraph that contains z and some of its predecessors such that $|\text{input}(F_z)| < K$. Hence, a K-feasible logic cone can be implemented by a single K-LUT.

The goal of technology mapping is to cover a design with K-feasible logic cones. Any node in the design would usually have a large number of K-feasible logic cones. Technology mapping selects the logic cone that maximizes the objective function of the mapping.

The K-feasible logic cones are obtained by finding the K-feasible cuts of the graph. A K-feasible cut $(X, \overline{X})$ for a node z is defined as a partition of the nodes in the subgraph rooted at z such that $z \in \overline{X}$ and the number of nodes in X that fanout to nodes in $\overline{X}$ is $\leq K$. Figure 4.21a shows a DAG with two 4-feasible cuts for the subgraph rooted at z. Using cut 2, the 4-feasible logic cones in Fig. 4.21b are obtained from the DAG in Fig. 4.21a. It can be noticed from Fig. 4.21b that the logic cones are obtained without duplicating any of the nodes among any of the logic cones.

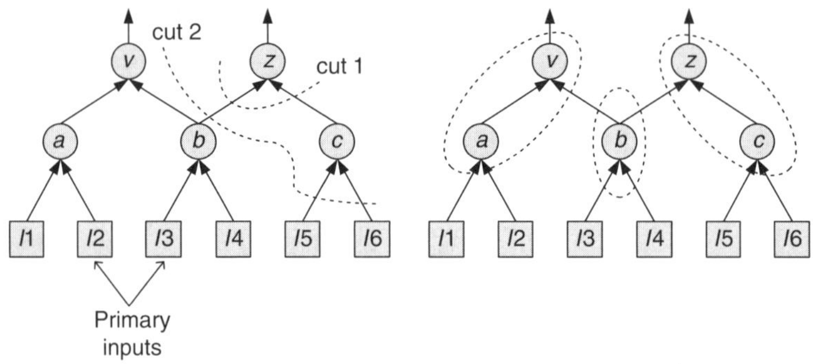

(a) A DAG with two 4-feasible cuts. (b) 4-feasible cones obtained from cut 2.

■ **FIGURE 4.21** Obtaining feasible logic cones from feasible cuts [122].

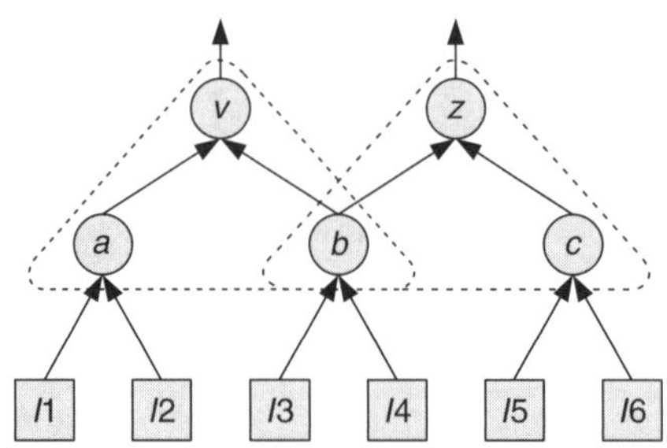

■ **FIGURE 4.22** Logic replication [122].

Logic replication is used whenever a LUT is used to implement a K-feasible cone that contains a node having a fanout outside that cone. Logic replication is mainly used to reduce the depth of the mapped circuit. Figure 4.22 depicts the same circuit in Fig. 4.21 divided into two 4-feasible logic cones using logic replication for node b. The circuit in Fig. 4.21b has a logic depth of two and uses three LUTs, whereas the implementation in Fig. 4.22 uses only two LUTs and has a logic depth of one.

By studying the implementation in Fig. 4.22, two observations can be made about the impact of logic replication on the routing resources: (1) connections between the replicated node and its fanouts may be covered by the LUTs, edges $b \to v$ and $b \to z$, thus reducing the use of the FPGA routing resources, and (2) the fanout of the nodes that fanin to the replicated node increases, edges $I3 \to b$ and $I4 \to b$. Hence, it can be concluded that replication can either end up increasing or decreasing the total wire length. However, in terms of power dissipation, it all depends on the switching activities of the connections that got covered and those which got created. If the net switching activities decreased, then the total power dissipation will decrease, and vice versa.

Anderson and Najm [122] developed a mapping algorithm for FPGAs that tries to minimize the total switching activities of the nets through logic replication. The algorithm is divided into three different phases. First, a set of K-feasible cuts is generated for each node in the network. Second, the algorithm calculates the cost of each cut and selects the one that results in the maximum power savings. Third, the algorithm evaluates the K-feasible logic cones that correspond to the optimum cut.

To find all the K-feasible cuts, the DAG is traversed starting from the primary inputs to primary outputs. As an example, consider node z in Fig. 4.22. Since the DAG is traversed from inputs to outputs, nodes b and c have already been traversed. Assume that b has two feasible K-feasible cuts C_{b1} and C_{b2}, whereas c has only one K-feasible cut C_c. C_{b1} and C_c can be merged to include node z, thus creating a new cut C_z, if: (1) $\text{Nodes}(C_z) = z \cup \text{Nodes}(C_{b1}) \cup \text{Nodes}(C_c)$, (2) $\text{support}(C_z) = \text{support}(C_{b1}) \cup \text{support}(C_c)$, and (3) $|\text{support}(C_z)| \leq K$, where $\text{support}(C_z)$ denotes the nodes that fanin to nodes in C_z. Traversing the DAG in a breadth-first approach using the same procedures would generate all the cuts of the DAG.

The authors [122] used a cost function that takes into consideration the depth $\text{DCost}(C_z)$, power $\text{PCost}(C_z)$, and logic replication $\text{RCost}(C_z)$ of cut C_z rooted at z. The total cost is given by

$$\text{Cost}(C_z) = \alpha \times \text{DCost}(C_z) + \beta \times \text{PCost}(C_z) + \gamma \times \text{DCost}(C_z), \quad (4.10)$$

where α, β, and γ are weights for each cost parameter. The depth cost DCost(C_z) is given by [122]

$$\text{DCost}(C_z) = 1 + \max_{v \in \text{support}(C_z)} \{\text{DCost}(\text{BestCut(v)})\}, \quad (4.11)$$

where BestCut(v) is the cut that results in the maximum cost for the subgraph rooted at v.

The power cost function in Eq. (4.10) is expressed as

$$\text{PCost}(C_z) = \sum_{v \in \text{support}(C_z)} [f_v + \text{PCost}(\text{BestCut(v)})] - \sum_{w \in \text{support}(C_z)} [f_w \times |\text{output}(w) \cap \text{Nodes}(C_z)|], \quad (4.12)$$

where f_x is the switching activity of the net driven by node x. The first summation in Eq. (4.12) represents the activities of the connections in the mapping of the subgraph rooted at z. The first term of that summation models the switching activities of nodes that fanout to a node in Nodes(C_z), which will need to be routed through the interconnect if Nodes(C_z) is implemented as a LUT in the final mapping. The second term in the first summation reflects the power cost of the mapping solutions rooted at each of the support nodes. The second summation term represents the sum of the fanout-weighted switching activity on the connections that have been captured inside a LUT if Nodes(C_z) is implemented as a LUT. For each node w in Nodes(C_z), this term counts the number of fanouts of w that are in Nodes(C_z) and multiplies this count by the activity of the signal driven by w.

To derive an expression for the logic replication cost in Eq. (4.10), Anderson and Najm [122] define the SlackWeight(z) of node z as

$$\text{SlackWeight}(z) = 1 + \kappa \times \left[\frac{\text{MaxSlack} - \text{Slack}(z)}{\text{MaxSlack}}\right], \quad (4.13)$$

where κ is a positive real number, MaxSlack is the maximum slack of all nodes in the DAG, and Slack(z) is the slack at node z. Slack(z) is calculated as the number of levels in the circuit Boolean network DAG by which the depth of node z may be increased, without increasing the overall depth of the DAG. The RCost(z) is formulated as

$$\text{RCost}(C_z) = \frac{1}{\text{SlackWeight}(z)} \times \sum_{v \in \text{RNodes}(C_z)} \left[\left(\sum_{u \in \text{input}(v)} f_u\right) - \lambda f_v \times \left|\text{output}(v) \cap \text{Nodes}(C_z)\right|\right],$$

where RNodes(z) is the set of nodes in C_z that fanout to a node outside C_z and λ is a constant set to 0.5.

The first summation in Eq. (4.14) is performed over the set of replicated nodes in C_z. For each replicated node, v, the activities of the signals driven by v fanins are summed to reflect the fact that replicating a node generally increases the fanout of its fanin nodes. The impact of logic replication on power dissipation depends on the activities of the signals driven by these fanin nodes and hence, RCost is increased in proportion to these activities. The second term in Eq. (4.14) reflects the fact that logic replication will result in the nets in its fanouts disappearing inside the LUT. Hence, RCost is decreased in proportion to the product of the number of captured connections and activity of these captured connections. RCost is inversely proportional to the SlackWeight to reduce the replication cost for the critical nodes whose depth in the mapping is likely to impact the overall depth of the mapped circuit.

Table 4.9 Power, Area, and Number of Connections Results for the Power-Aware Technology Mapping Technique [122]

Circuit	Depth	Power Savings (%)	Area Reduction (%)	No. of Connections Reduction (%)
C3540	15	11	9	5
C5315	11	2	3	4
alu4	7	20	11	11
apex1	7	14	0	0
apex2	8	15	8	7
apex3	6	19	10	9
ex5p	7	26	−3	−3
apex5	5	5	5	3
cordic	9	8	4	6
cps	5	16	6	5
dalu	6	16	3	7
des	7	6	1	2
Average for these circuits		13	5	5
Average across 29 circuits		14	5	6

In the study by Anderson and Najm [122], the mapping cost formulation in Eq. (4.10) is used to drive the FlowMap algorithm proposed by Cong and Ding [119]. Table 4.9 lists a comparison between the power, area, number of connections used in FlowMap, and the power-aware technology mapping techniques proposed by Anderson and Najm [122]. The results in Table 4.9 suggest that the power-aware technology mapping algorithm always results in

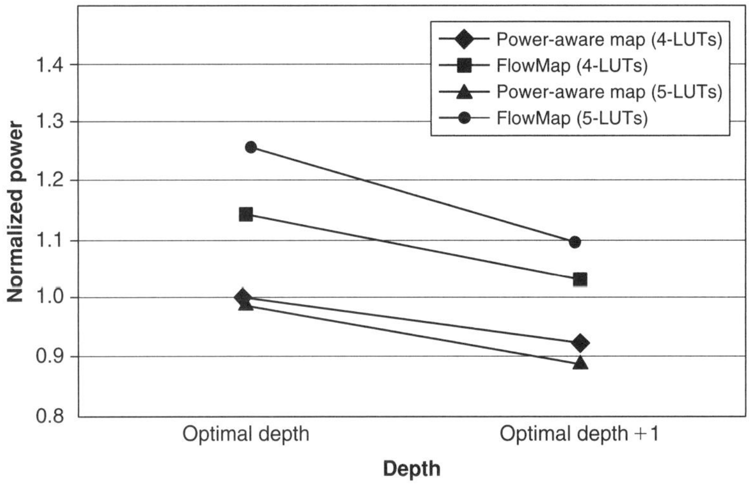

FIGURE 4.23 Power savings with the logic depth and LUT size [122].

power savings over the conventional FlowMap algorithm without adding any penalties on the total area of or the number of connections in the design. The average power savings achieved by the power-aware technology mapping algorithm is almost 16%. The only design that experimented an increase in the area and number of connections is the ex5p benchmark.

Figure 4.23 plots the power savings achieved by the power-aware technology mapping algorithm as the depth of the circuit is increased by 1 from the optimal depth. From Fig. 4.23, it can be shown that increasing the depth of the circuit by only 1 from the optimal depth would result in almost 8–10% power savings using the power-aware technology mapping. Moreover, for the FlowMap results, increasing the depth by 1 would result in a significant reduction in the power dissipation, yet the power dissipated is still higher than that dissipated by a design implemented using the power-aware technology mapping algorithm at optimal depth. These results show the significance of the power savings achieved by the power-aware technology mapping algorithm. Moreover, mapping to

5-LUT architectures results in slight power savings while using the power-aware technology mapper, while FlowMap results in an increase in the power dissipation.

4.3.1.2 *Energy-Aware Technology Mapping (EMap)*

Lamoureux and Wilton [123] proposed a new energy-aware technology mapper (EMap) that tries to minimize the total wire length during mapping rather than to minimize the wire length of the highly active nets as in the study by Anderson and Najm [122]. As a result, EMap tries to minimize the number of replicated nodes in the circuit. The EMap algorithm is composed of three phases as shown in Algorithm 4.4. The first phase generates all the K-feasible cuts for each node in the circuit adopting the same techniques used in the study by Cong et al [16] by processing all the circuit nodes in a topological order. While processing each node, a label is added to every node to represent its optimal depth.

In the second phase of the EMap algorithm, the cuts generated for each node in the first phase are processed in reverse topological order. If the node has zero slack, the cut that results in the optimal depth is selected; otherwise, the cut that has the minimum cost is selected. The cost of the cut $(X_v, \overline{X_v})$ is evaluated as

$$\mathrm{Cost}(X_v, \overline{X_v}) = \frac{1 + |\mathrm{rooted}(\overline{X_v})|}{1 + |\overline{X_v}| - |\mathrm{rooted}(\overline{X_v})|} \times \sum_{u \in \mathrm{input}(\overline{X_v})} \frac{\mathrm{weight}(u) \times \left(1 + \lambda \times \mathrm{act}(u)\right)}{|\mathrm{output}(u)|}, \tag{4.14}$$

where $\overline{X_v}$ is the set of nodes inside the LUT that correspond to the cut $(X_v, \overline{X_v})$, $\mathrm{rooted}(\overline{X_v})$ is the set of nodes in $\overline{X_v}$ that have been labeled as root nodes, $\mathrm{weight}(u)$ is a binary variable that is set to zero if node u is labeled as a root node of a LUT and one, otherwise, $\mathrm{act}(u)$ is an

Algorithm 4.4 EMap algorithm pseudocode [123]

```
Phase 1:
for each node v ∈ N do
  enumerate_K_feasible_cuts(v, K)
end for
for each node v ∈ N do
  label(v) = compute_label(v)
  if ((v ∈ primary_input(N)) || (v ∈ primary_output(N))) then
    rooted(v) = TRUE
  else
    rooted(v) = FALSE
  end if
end for
D_opt = max {label(v)|v ∈ N}
for each node v ∈ N do
  latest(v) = D_opt
  slack(v) = latest(v) − label(v)
end for

Phase 2:
for each node v ∈ N do
  if rooted(v) == TRUE then
    if slack(v) > 0 then
      (X_v, X̄_v) = choose_cut(K-feasible_cut(v));
    else
      (X_v, X̄_v) = choose_cut(min_height_K-feasible_cut(v));
    end if
    for each node u ∈ input(X_v, X̄_v) do
      rooted(u) = TRUE
      latest(u) = min (latest(u), latest(v) − 1)
      slack(u) = latest(u) − label(u)
    end for
  end if
end for

Phase 3:
for_LUT_network(N)
```

Table 4.10 Energy, Area, and Number of Connections Results for EMap [123] Compared to FlowMap

LUT Size	Energy Savings (%)	Area Reduction (%)	No. of Connections Reduction (%)
4	15.9	15.83	16.16
5	18.18	18.6	19.46
6	18.15	16.03	18.16

estimation of the switching activity of the net driven by u, λ is a parameter used to control the relative importance of the activity factor, and output(u) is the set of nodes that are fanouts of node u. The first part of the cost function in Eq. (4.20) minimizes node duplications by increasing the cost of cuts with root nodes. The second part in Eq. (4.20) favors cuts that have fewer inputs, since they would have lower connection activities.

Table 4.10 lists the results for the energy, area, and number of connections savings achieved by EMap [123] compared to that of the study by Cong et al [16]. The average energy savings achieved by EMap ranges from 15 to 18%. Comparing the results of EMap [123] with those of the power-aware technology mapping proposed by Anderson and Najm [122] is a bit difficult since both works used different power models to estimate the power savings achieved.

4.3.2 Power Reduction Techniques during Clustering

In modern clustered FPGA architectures, several LUTs and flip-flops are grouped together to form one cluster. Interconnects within the cluster are faster and dissipate less energy than connections between clusters. Traditionally, clustering algorithms use area minimization, delay minimization, and/or routability maximization as objective functions. Hence, clustering algorithms aim to (1) pack clusters to their maximum capacity, (2) cluster LUTs on the critical path together, and/or (3) minimize the number of inputs per cluster.

Intuitively, we would expect clustering to be more effective than technology mapping at reducing power, since clusters are typically larger (commercial parts contain as many as 10 LUTs per cluster). On the other hand, encapsulating high activity nodes into clusters does not eliminate these nodes entirely, as it does in technology mapping. An interconnection between LUTs within a cluster still requires a connection; however, the capacitance of this intracluster connection is much smaller than the capacitance of the intercluster connections.

Lamoureux and Wilton [123] proposed a clustering algorithm that targets power minimization, in addition to the area, delay, and routability objectives. The technique is based on the T-VPack tool [18, 27], explained earlier in Chapter 1. The power-aware clustering techniques achieve power reductions by modifying the Attraction(B) function between each logic block and the cluster, which is given in Eq. (1.7). The Attraction(B) function is modified according to

$$\begin{aligned}\text{Attraction(B)} &= \lambda \times \text{Criticality(B)}\\ &+ (1-\lambda) \times \left[(1-\beta) \times \frac{\sum \text{weight}(i) | i \in \text{Nets(B)} \cap \text{Nets(C)}}{\text{MaxNets}}\right.\\ &\left.+ \beta \times \frac{\sum \text{Activity}(i) | i \in \text{Nets(B)} \cap \text{Nets(C)}}{\text{MaxNets} \times \text{Activity}_{avg}}\right], \end{aligned} \tag{4.15}$$

where Activity(i) is the switching activity of net i, Activity_{avg} is the average switching activity of all the nets in the circuit, β is a constant used to give more importance for one part of the relationship over the other, and weight(i) is a function used to model complete net encapsulation inside a cluster. If net i connects block B to another block in cluster c, then weight(i) can be either 1 or 2; otherwise,

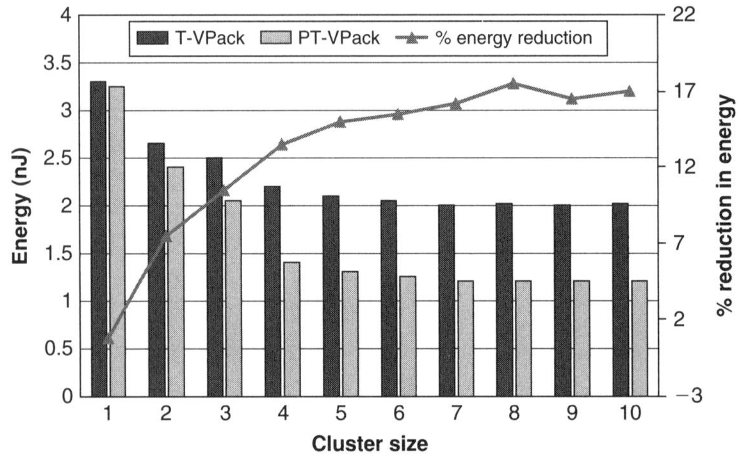

■ **FIGURE 4.24** Energy savings achieved by the power-aware clustering technique [123].

it is set to 0. If net i connects B to C and it has less than four pins and has not been connected to any other cluster before, weight(i) is set to 2. This gives more incentive to complete net encapsulation within a cluster. The third term in Eq. (4.15) minimizes the switching activity of connections between clusters by forcing highly active nets to be engulfed inside the cluster.

Figure 4.24 plots the average energy over a large number of FPGA benchmarks for the T-VPack clustering technique and power-aware clustering technique (PT-VPack) proposed in an earlier study [123]. It can be observed from Fig. 4.24 that increasing the cluster size results in more energy savings due to the PT-VPack technique. This is mainly due to the fact that bigger clusters means more nets can be fully engulfed within a cluster, hence, more energy savings. At a cluster size of four, the average energy savings is approximately 12.6%.

4.3.3 **Power Reduction Techniques during Placement and Routing**

The placement and routing stages of the FPGA design are concerned with associating each cluster and net to a physical location on the FPGA. Most power efficient FPGA placement tools try to place clusters that are connected by highly active nets close to each to reduce the total capacitance of their connections [123–125].

4.3.3.1 ***Power-Aware Placement Technique***

Lamoureux and Wilton [123] proposed a power-aware placement technique based on the VPR placement algorithm [28]. The technique adds a power cost function to the cost function used by VPR. VPR uses a cost function that consists of two parts: a wire-length cost (WiringCost) and a timing cost (TimingCost). The wire-length cost function is based on bounding box wire-length calculations and is given by

$$\text{WiringCost} = \sum_{i=1}^{N_{\text{nets}}} q(i) \times [\text{bb}_x(i) + \text{bb}_y(i)], \tag{4.16}$$

where N_{nets} is the total number of nets, $\text{bb}_x(i)$ and $\text{bb}_y(i)$ are the x and y bounding box dimensions of net i, and $q(i)$ is a scaling factor to increase the accuracy of the wire-length estimation.

The TimingCost is given by

$$\text{TimingCost} = \sum_{\forall i,j \in \text{circuit}} \text{Delay}(i,j) \times \text{Criticality}(i,j)^{\text{CE}}, \tag{4.17}$$

where $\text{Delay}(i,j)$ is the delay of the net connecting i and j, $\text{Criticality}(i,j)$ is a measure of how critical the net is, and CE is a constant. Hence, the cost function used by VPR is given by

$$\Delta C = \lambda \times \frac{\Delta\text{TimingCost}}{\text{PreviousTimingCost}} + (1-\lambda) \times \frac{\Delta\text{WiringCost}}{\text{PreviousWiringCost}}, \tag{4.18}$$

where PreviousTimingCost and PreviousWiringCost are the cost functions evaluated in the previous simulated annealing temperature iteration and λ is a constant used to give importance of one cost function over the other.

The PowerCost introduced in the study by Lamoureux and Wilton [123] is given by

$$\text{PowerCost} = \sum_{i=1}^{N_{\text{nets}}} q(i) \times [\text{bb}_x(i) + \text{bb}_y(i)] \times \text{Activity}(i), \tag{4.19}$$

where Activity(i) is a measure of the switching activity of net i. Hence, the cost function used [123] is given by

$$\Delta C = (1-\gamma) \times \left[\lambda \times \frac{\Delta\text{TimingCost}}{\text{PreviousTimingCost}} + (1-\lambda) \times \frac{\Delta\text{WiringCost}}{\text{PreviousWiringCost}}\right] + \gamma \times \frac{\Delta\text{PowerCost}}{\text{PreviousPowerCost}}, \tag{4.20}$$

where γ is a parameter used to give importance to the PowerCost over the other cost functions. Applying the power-aware placement algorithm in reference [123] results in an average energy savings of only 3%; however, the delay of the critical path increases by approximately 4%.

4.3.3.2 *Power-Aware Routing Technique*

To reduce dynamic power dissipation during routing, the routing algorithm needs to make sure that the nets with high switching activities be kept as short as possible. Lamoureux and Wilton [123] achieve this goal by adding a term to the routing cost function that depends on the net switching activity. The power-aware routing algorithm [123] uses the following formulation instead of the one in Eq. (1.11):

$$\begin{aligned}\text{Cost}(n) = {} & \text{Crit}(i,j) \times \text{delay}(n, \text{topology}) \\ & + \left[1 - \text{Crit}(i,j)\right] \times \Big[\text{ActCrit}(i) \times \text{Cap}(n) \\ & + \big(1 - \text{ActCrit}(i)\big) \times b(n) \times h(n) \times p(n)\Big],\end{aligned} \tag{4.21}$$

where $\text{Cap}(n)$ is the capacitance of net n and $\text{ActCrit}(i)$ is the activity criticality and is given by

$$\text{ActCrit}(i) = \min\left\{\frac{\text{Activity}(i)}{\text{MaxActivity}}, \text{MaxActCrit}\right\},$$

where $\text{Activity}(i)$ is the switching activity in net i, MaxActivity is the maximum switching activity of all the nets, and MaxActCrit is the maximum activity criticality that any net can have. Applying the power-aware routing algorithm in reference [123] results in an average energy savings of only 2.6%; however, the delay of the critical path increases by approximately 3.8%.

4.3.3.3 *Power-Aware Placement and Routing Techniques in Commercial FPGAs*

Gupta et al [124] and Vorwerk et al [125] present a low-power placer and router for Xilinx and Actel FPGAs, respectively. The placement and routing techniques try to minimize the length of the nets with high switching activities but without adding any performance penalties to critical nets. Critical nets are routed in a timing-driven approach rather than a power-driven approach. The authors recorded an average power savings of 8.6%, whereas the performance degradation was in the range of 3–4% [124]. The average power savings was 13%, whereas the performance loss was limited to 1% [125]. The main reason for the difference in the results between the techniques proposed in the above studies [124, 125] is the difference between the FPGA architecture the algorithms were applied to.

Chapter 5

Leakage Power Reduction in FPGAs Using MTCMOS Techniques

This chapter presents supply gating techniques in FPGAs through the use of multithreshold CMOS (MTCMOS) approaches for subthreshold leakage power reduction [126–128]. A modified FPGA architecture with sleep transistors is discussed and the CAD algorithms needed to benefit from the architecture changes are introduced. Specifically, a new activity profiling phase is introduced in the CAD flow to identify the blocks that exhibit similar idleness to collectively turn them OFF during their idle times. Moreover, new packing techniques are presented to pack those blocks with similar activity profiles together to easily turn them OFF.

5.1 INTRODUCTION

In FPGA designs, leakage power reduction has been overshadowed by performance improvements and dynamic power minimization techniques. However, recently, leakage power started to gain increased attention by both FPGA circuits and CAD designers. The leakage power dissipation problem is more crucial in FPGAs compared to custom ASIC designs because of the unutilized resources in FPGAs. On average, the percentage utilization of resources in FPGAs is approximately 60% [129]. Thus, almost 40% of the FPGA consumes standby leakage power without delivering useful output. Moreover, FPGAs used in wireless applications can go into idle mode for long periods of time [130]. In such designs, even the utilized resources need to be forced into a low-power (standby) mode during their idle periods to save leakage power.

One of the most popular techniques used in leakage power reduction in ASIC designs is multithreshold CMOS (MTCMOS) [131, 132]. In an MTCMOS implementation, a high-V_{TH} (HVT) device called the *sleep transistor* connects the pull-down network using low-V_{TH} (LVT) devices of a circuit to the ground, as shown in Fig. 5.1a. When the sleep transistor is turned OFF, the circuit subthreshold leakage current is limited to that of the sleep transistor, which is significantly low. Hence, the circuit benefits from the high performance of the LVT pull-down network when the sleep transistor is turned ON, while limiting the circuit subthreshold leakage current when the sleep transistor is turned OFF.

The sleep transistor acts as a small finite resistance R to the ground when the SLEEP signal is *high* with a finite small voltage at the virtual ground rail V_x, as shown in Fig. 5.1b. However, the sleep transistor resistance R incurs a performance penalty because the driving potential of the circuit is reduced to $V_{DD} - V_x$ [130, 133]. When the SLEEP signal is *low*, the circuit goes into a standby mode with the voltage at V_x rising to a voltage between 0 and V_{DD}, with the sleep transistor acting as a very high resistance,

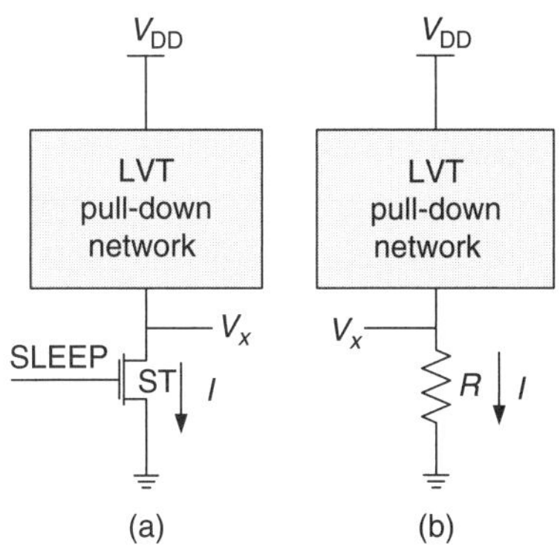

■ **FIGURE 5.1** MTCMOS architecture. (a) General MTCMOS architecture, (b) equivalent ST circuit in the active mode.

thus reducing the standby subthreshold leakage current considerably.

In FPGAs, sleep transistors can reduce subthreshold leakage by (1) *permanently* powering down the unutilized parts of the chip per configuration, (2) *dynamically* turning ON and OFF the utilized parts of the chip depending on their activity, and (3) powering down all of (or a large part of) the FPGA during the design idle time.

In this chapter, the MTCMOS technique is used in FPGA design and the changes needed at the CAD level are developed to take full advantage of the technique in maximizing the leakage savings. These changes are integrated into the academic versatile place and route (VPR) flow [27]. A flowchart of a typical VPR CAD flow is shown in Fig. 5.2a and a flowchart of the proposed modifications is shown in Fig. 5.2b. In Fig. 5.2b, a new stage is

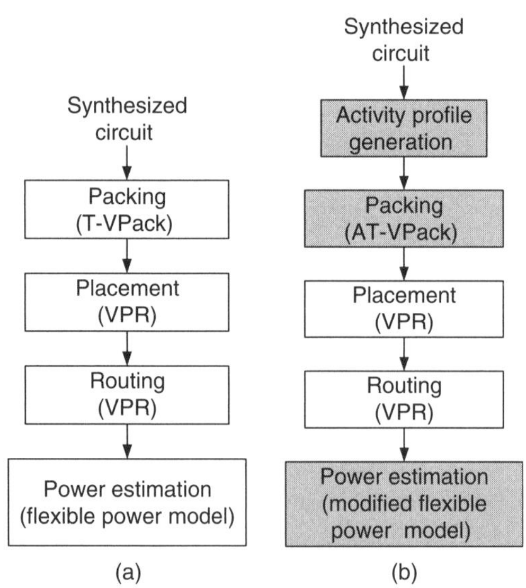

■ **FIGURE 5.2** FPGA CAD flowchart. (a) Conventional VPR flowchart. (b) Proposed CAD flowchart integrated in the VPR flow.

added to the CAD flow, the activity generation phase, in which the design is analyzed to identify the logic blocks that exhibit similar *activity profiles*. Blocks with similar activity profiles are forced into a standby mode together. The activity profiles generated by the activity generation algorithms are then integrated into the T-VPack algorithm to result in the activity T-VPack algorithm (AT-VPack), as shown in Fig. 5.2b. A modified power model that takes into consideration the proposed changes in the FPGA architecture is used to properly calculate the power savings from the proposed MTCMOS FPGA architecture.

5.2 MTCMOS FPGA ARCHITECTURE

The conventional hierarchical FPGA architecture, adopted by most modern FPGAs, uses logic blocks, which are conventionally made of a 4-input look-up table (LUT), a flip-flop, and a 2:1 multiplexer, as shown in Fig. 5.3. Several logic blocks are further grouped together to form a cluster of logic blocks. Inside each cluster, the logic blocks are connected using the local routing resources, whereas the clusters are connected using the global routing resources.

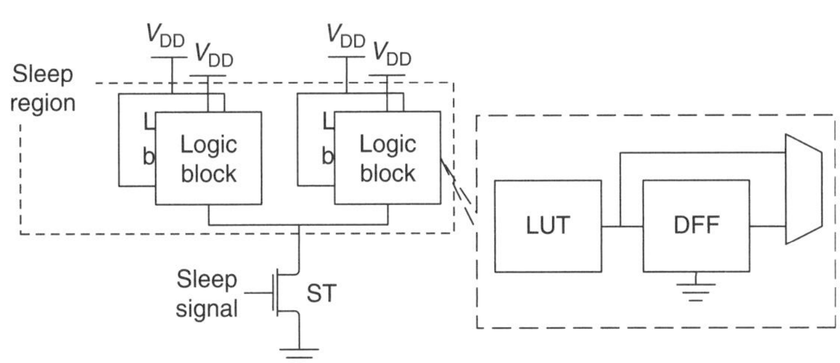

■ **FIGURE 5.3** MTCMOS FPGA architecture. The logic blocks connected to one sleep transistor are called the *sleep region*.

The MTCMOS FPGA architecture proposed by Hassan et al [127] follows the broad guidelines of the hierarchical architecture; however, every N clusters are connected to the ground through one sleep transistor, as shown in Fig. 5.3. Moreover, the latches in each cluster are used to retain the value of the logic block outputs when they enter the sleep mode. Thus, they are not connected to the sleep transistors. The logic blocks served by one sleep transistor are called the *sleep region*. It should be noted that the sleep transistors are not confined to the logic resources of the FPGA but are applied to the routing resources of the fabric as well.

Each sleep transistor is controlled by a SLEEP signal. Deactivating the SLEEP signal forces the N clusters in the corresponding sleep region into low-power mode during their inactive periods. Before entering the sleep mode, the output of each logic block is stored in the latch, so it can be recovered when the sleep region wakes up again. The SLEEP signals of the unutilized, whether logic or routing, resources of the FPGA are kept deactivated at all times to turn them permanently OFF.

The SLEEP signals are generated dynamically during the device runtime using the partial reconfiguration logic available in modern FPGAs [134, 135], thus providing minimum area overhead. The SLEEP signals can be generated if the application of the design is well known in advance. For example, if the design is used to implement an MPEG decoder, then the sequence of operations to be executed is known in advance as well as the statistics of each signal, which can then be used to generate the SLEEP signals as will be explained later in Section 5.4. This is a very interesting point since the majority of the FPGA applications are indeed dedicated ones where the application is well known in advance. However, if the design application is a general one, earlier works formulated a methodology for predicting the statistics of the design signals in a methodology similar to branch prediction methodologies [136].

The number of clusters that can fit in one sleep region is determined by (1) the size of the sleep transistor, which in turn corresponds to the maximum performance loss allowed; (2) leakage power savings; (3) area overhead permitted in the design due to sleep transistors; and (4) the maximum permitted ground bounce on the virtual ground lines. For the same performance penalty, large sleep regions use larger, but fewer in number, sleep transistors. As a result, the control circuitry needed to generate the SLEEP signals is typically less complex, consumes less power, and occupies a smaller area compared to the small sleep regions. However, large sleep regions have limited leakage power savings capability due to the use of large sleep transistors, which sink larger subthreshold leakage current. Moreover, large sleep regions suffer from a smaller selectivity in turning OFF idle clusters, thus reducing their resulting leakage power savings. Hence, the optimum granularity is set based on a compromise between the area overhead and the required leakage power savings. In reference [137], the authors concluded that the optimum granularity ranges from four to eight logic blocks.

A diagram of the proposed FPGA fabric is shown in Fig. 5.4, where the sleep transistors are prefabricated with a fixed size in the FPGA fabric. It was shown in the study by Kosonocky et al [138] that such a placement provides the minimum area overhead while ensuring full connectivity between the sleep transistors and the logic blocks. Moreover, the SLEEP control signals for each sleep transistor are hardwired during the FPGA fabrication. The virtual ground VGND line is used to connect the pull-down networks of the logic blocks to the sleep transistor, as shown in Fig. 5.4. The VGND lines are hardwired to their corresponding sleep transistors. Several research works

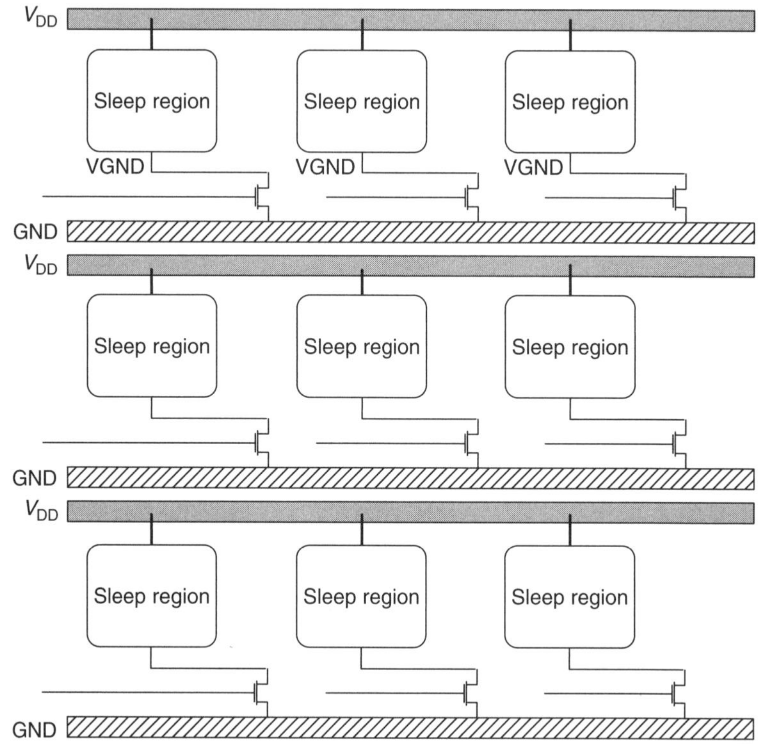

■ **FIGURE 5.4** MTCMOS-based FPGA fabric with sleep transistors.

proposed optimum layouts for the sleep transistors to provide the minimum area overhead [139], and the average area overhead of MTCMOS architectures with fine granularity (from four to eight logic blocks) in FPGAs is reported to be approximately 5% [140, 141].

It should be noted that there are two approaches for sleep transistor implementations: header or footer devices. Header devices use a PMOS sleep transistor to block the path from the supply rail, whereas the footer approach uses an NMOS to block the path to the ground, as shown in Fig. 5.5. The PMOS header approach has the disadvantage of incurring a large-area penalty compared to the

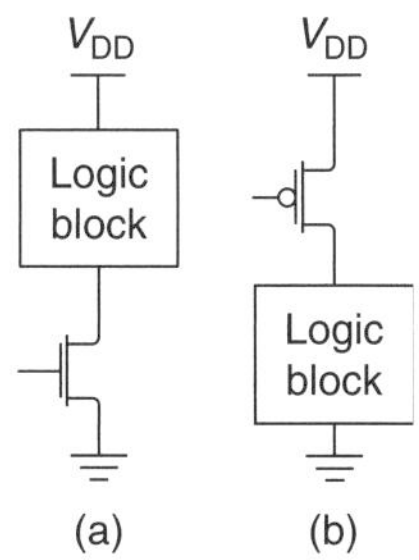

■ **FIGURE 5.5** Sleep transistor implementations. (a) NMOS footer. (b) PMOS header.

NMOS footer approach. This is mainly because of the lower drive current of PMOS devices due to the lower mobility of holes compared to electrons. As a result, to have the same performance penalty due to sleep transistors, PMOS headers with larger areas are needed. Consequently, only footer NMOS sleep transistors are used [127].

Typically, there are two sleep transistor architectures: local and global sleep transistors. Local, or distributed, sleep transistors are placed at the local block level, where the local block is defined as a part of the circuit that can be independently idle. On the other hand, a global sleep transistor architecture uses a single sleep transistor for a large circuit block that includes several local blocks. In reference [127], a local sleep transistor architecture is adopted for the following reasons: (1) the VGND lines are short enough to be treated as local connections, and hence, there is no need to fabricate them using wide metal lines like V_{DD} and GND rails; (2) the routing complexity of the VGND lines is significantly easy in local sleep transistor architectures compared to the global architecture; and (3) local sleep transistors provide less routing overhead in terms of the criticality of the sleep signals, better noise margins, and higher turn OFF flexibility, thus higher power savings. However, the control systems for local sleep transistor architecture are more complicated.

5.3 SLEEP TRANSISTOR DESIGN AND DISCHARGE CURRENT PROCESSING

In this section, several issues related to the sleep region are discussed. First, the design problem of the sleep transistor is introduced and a formulation for the transistor size is presented in terms of the total discharge current of the sleep region. Second, two methods for total sleep region discharge current calculations are proposed. The first one is a modified version of the mutually exclusive discharge current algorithm proposed by Anis et al [133]. The second method is a newly proposed algorithm that considers the logic function implemented by the logic blocks.

5.3.1 Sleep Transistor Sizing

The proper sizing of the sleep transistor is crucial to achieve the maximum subthreshold leakage power savings without incurring large performance and area penalties, as explained in Section 5.2. While the delay penalty is inversely proportional to the width of the sleep transistor, a large sleep transistor results in a large subthreshold leakage current and higher parasitic capacitances, which results in high dynamic power dissipation during the switching of the sleep transistor. Moreover, a large sleep transistor consumes a larger part of the total chip area. The first step in sizing the sleep transistor is to formulate the delay penalty experienced by the FPGA circuitry due to the sleep transistors.

The delay of a CMOS gate without any sleep transistors t_d is expressed as [131, 142]

$$t_d \propto \frac{C_L V_{DD}}{(V_{DD} - V_{TH_l})^{\alpha}}, \tag{5.1}$$

where C_L is the gate load capacitance, V_{TH_l} is the threshold voltage of the circuit LVT, and α is the velocity saturation index. The delay of the same gate in the presence of a sleep transistor $t_{\text{d,sleep}}$ is expressed as [133]

$$t_{\text{d,sleep}} \propto \frac{C_{\text{L}} V_{\text{DD}}}{(V_{\text{DD}} - V_x - V_{\text{TH}_\text{l}})^{\alpha}}, \tag{5.2}$$

where V_x is virtual ground rail voltage, as shown in Fig. 5.1a.

To balance between the performance penalty and power savings, the maximum allowable performance loss should be limited to a predefined value. Let the ratio between t_{d} and $t_{\text{d,sleep}}$ be given by

$$\frac{t_{\text{d,sleep}} - t_{\text{d}}}{t_{\text{d,sleep}}} = x, \tag{5.3}$$

where x is the performance loss due to sleep transistors. For simplicity, assume that α can be approximated to be equal to 1 [133]. Therefore, substituting with Eqs. (5.1) and (5.2) into (5.3) yields

$$V_x = x \times (V_{\text{DD}} - V_{\text{TH}_\text{l}}). \tag{5.4}$$

When the sleep transistor is turned ON, it will operate in the linear mode of operation, as explained earlier. Using the square law for the MOS device current, the drain to source current flowing through the sleep transistor I_{sleep}, i.e., discharge current, can be approximated by

$$I_{\text{sleep}} = \mu_n C_{\text{ox}} \left(\frac{W}{L}\right)_{\text{sleep}} \left[(V_{\text{DD}} - V_{\text{TH}_\text{h}}) \times V_x - \frac{V_x^2}{2}\right], \tag{5.5}$$

where μ_n is the device mobility, C_{ox} is the oxide thickness, W and L are the device width and length, respectively, and V_{TH_h} is the threshold voltage of the sleep transistor, which is an HVT device. Substituting with the expression of V_x given in Eq. (5.4) into the value of I_{sleep} in Eq. (5.5) and rearranging the relationship results in [133]

$$\left.\frac{W}{L}\right|_{\text{sleep}} = \frac{I_{\text{sleep}}}{x\mu_n C_{\text{ox}}(V_{\text{DD}} - V_{\text{TH}_\text{l}})(V_{\text{DD}} - V_{\text{TH}_\text{h}})}. \tag{5.6}$$

The parameters in Eq. (5.6) are all technology parameters except for the speed penalty x and the maximum discharge current allowed through the sleep transistor I_{sleep}. In most earlier works, x has been set to a constant value, usually 5% [133]. As a result, all the circuit paths will experience a fixed speed degradation. However, in the study by Hassan et al [127], two possibilities are explored: setting x to a fixed value as well as using variable speed penalties to improve the final design performance, as will be explained in Section 5.5.

The next step in finding the proper size of the sleep transistor is to compute the sleep region maximum discharge current I_{sleep}. It should be noted that reference [127] uses footer devices, i.e., NMOS devices, as sleep transistors. Hence, the main criterion that controls the sizing of footer devices is the discharge current of the pull-down network. The charging current flows through the pull-up circuit and the sleep transistor is not involved in this process, hence, the charging time is not affected.

The worst-case maximum value for I_{sleep} is the sum of discharge currents of all the logic blocks inside the sleep region. Since all the logic blocks in FPGAs are identical, then the values of their discharge currents would be equal. As a result, the value of I_{sleep} would be expressed as

$$I_{\text{sleep}} = I_{\text{discharge}} \times N, \tag{5.7}$$

where $I_{\text{discharge}}$ is the discharge current of one logic block and N is the granularity of the sleep region, i.e., the number of logic blocks in one sleep region.

However, this is more of an upper bound on the value of I_{sleep} due to two factors: (1) the delays of the logic blocks are finite and (2) not all the logic blocks inside the sleep region will discharge simultaneously. The choice and computation of the discharge current I_{sleep} inside the sleep region are explained in the following subsections. Selecting a value for I_{sleep} depends on the allowable number of logic blocks in each sleep region. To find the optimum value of I_{sleep} to be used, the discharge current of each cluster placed using the conventional VPR tool is calculated for several FPGA benchmarks. It was found that the value of the discharge current of all the clusters is usually less than 75% of the worst-case discharge current, which is therefore used in reference [127]. However, it should be noted that the sum of discharge currents of all the logic blocks inside the cluster must not exceed the value of I_{sleep}, or else the sleep region will experience a bigger speed penalty than that used to evaluate $\left(\frac{W}{L}\right)_{\text{sleep}}$ in Eq. (5.6).

5.3.2 Mutually Exclusive Discharge Current Processing

The mutually exclusive discharge current processing technique was first proposed by Anis et al [133] for standard cell MTCMOS design. This technique makes use of the finite delays of each gate to provide a sequence of discharge current patterns inside each sleep region. The discharge current of any logic gate is represented using a symmetric triangular approximation, as shown in Fig. 5.6b. Due to the finite delay of A in Fig. 5.6a and the dependence of B on the inputs to A, B will not start discharging before the discharge current of A reaches its peak [133]. In this case, A and B are said to be mutually exclusive in their discharge current, since they are not going to discharge simultaneously.

In Fig. 5.6a, two parameters characterize the discharge current of each gate: the maximum value the discharge current can reach I_i and the time it takes for the discharge current to reach its peak T_i, as shown in Fig. 5.6b.

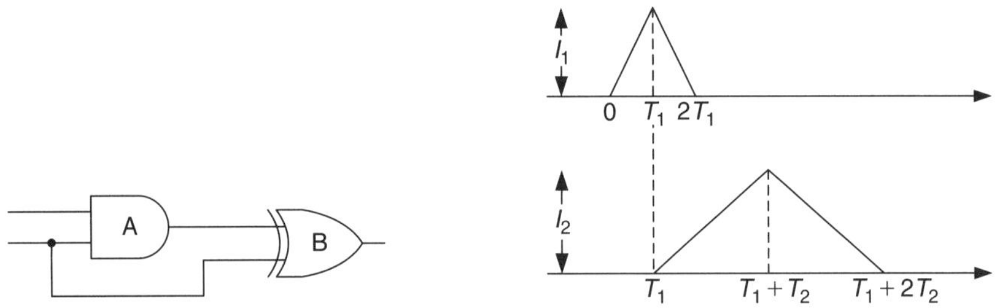

(a) A circuit example for the timing diagram. (b) Timing diagram for mutually exclusive logic gates.

■ FIGURE 5.6 Mutually exclusive discharge current processing.

The values of these two parameters depend on the type of the gate, since every gate would have a different delay and maximum discharge current, and the fanout, increasing the gate fanout slows down the discharge by decreasing the value of I_i and increasing T_i. Hence, to use this technique, all gates in the design library are characterized initially by simulating their discharge currents under all possible loading scenarios using HSpice.

Applying this technique for discharge current processing in FPGAs is much simpler than the standard cell case due to the regularity of FPGAs. First, all FPGA logic blocks are identical, since a k-input logic block can implement any k-input logic function. As a result, only one circuit is characterized using HSpice. Second, the loading effect in FPGAs is very uniform due to the use of routing switches. Hence, there is a very limited number of loading scenarios that can be experienced. These two facts decrease the number of HSpice simulations needed to characterize the logic gates significantly. As an approximation, Hassan et al [127] assumes that the discharge current patterns, in terms of peak value and duration, are the same for all the logic blocks in the design.

If the small circuit example in Fig. 5.6a is implemented in an FPGA, the discharge current of these two logic blocks will be represented as shown in Fig. 5.7a. It can be noticed

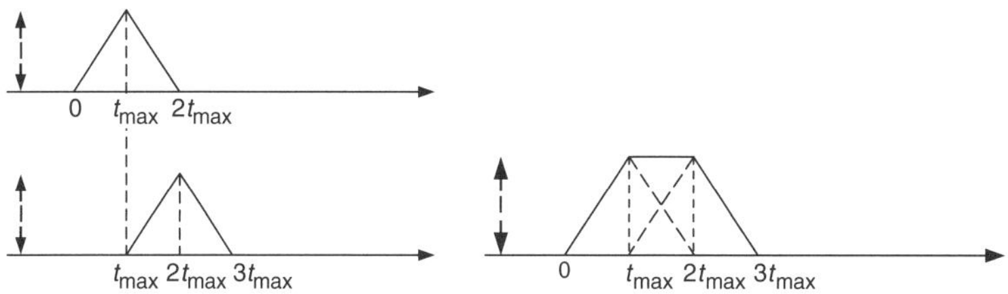

(a) FPGA timing diagram of the circuit in Fig. 5.6a.

(b) Summation of discharge currents.

■ **FIGURE 5.7** Mutually exclusive discharge current processing.

in Fig. 5.7a that the discharge currents of both A and B are identical to reflect the fact that FPGA logic blocks are identical. These two discharge currents can be summed in a vector manner to result in the total discharge current for these two logic blocks, as shown in Fig. 5.7b. Hence, if these two logic blocks are placed in one sleep region, then the maximum discharge current that this sleep region will ever experience is only equal to $I_{discharge}$ of one logic block. This proves the worst-case value of I_{sleep} given in Eq. (5.7) is not always needed for the logic blocks inside the sleep region.

5.3.3 Logic-Based Discharge Current Processing

Earlier MTCMOS works adopted a worst-case discharge current processing algorithm by assuming that whenever a logic block A discharges, all of its outputs will start discharging after the discharge current of A reaches its maximum [133], as shown in Fig. 5.8b. However, the discharge of the fanout logic blocks of A will depend on the logic they implement. Therefore, a more efficient current processing algorithm has to include the probability of the circuit actually discharging based on the logic function implemented by the circuit.

For example, considering the small circuit in Fig. 5.8a, assume that B implements the following logic function: $b = \overline{a} + az$. Hence, whenever the output of A goes low, the output of B will always go high. Consequently, A and B

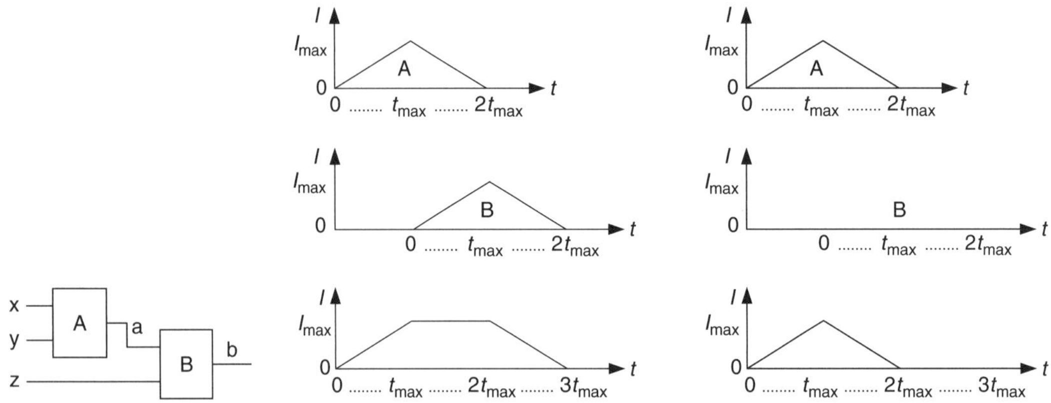

(a) Small circuit example.

(b) Logic-based discharge current processing for nonmutually exclusive logic blocks.

(c) Logic-based discharge current processing for mutually exclusive logic blocks.

■ **FIGURE 5.8** Linear vector approximation of discharge current and logic-based current vectors summation.

are mutually exclusive in their discharge. As a result, the total discharge current of these two logic blocks would only be equivalent to one of them, as shown in Fig. 5.8c. As a result, adding block B to the sleep region that contains A comes at no expense in terms of the discharge current of the sleep region—hence, speed penalty. This property will give the packing algorithms, which will be introduced later on in Section 5.5, more flexibility in packing logic blocks in the same sleep region without violating the maximum discharge current constraint of the sleep region. This new technique used in calculating the total discharge current inside a sleep region is called *logic-based discharge current processing.*

5.3.4 Topological Sorting and Discharge Current Addition

Topological sorting is used to properly align the current vectors in the sleep region to find the total discharge current [127]. The topological sorting algorithm encounters three different types of sleep regions: a combinational

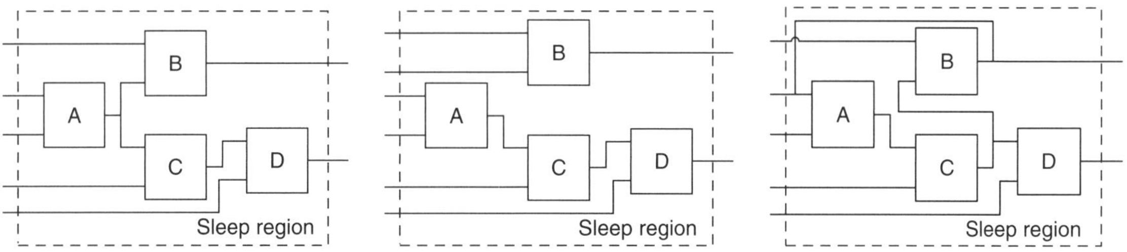

(a) Combinational connected. (b) Combinational with unconnected blocks. (c) Sequential with loops.

■ **FIGURE 5.9** Different types of sleep regions.

sleep region where each logic block shares at least one net with any other logic block in the sleep region (Fig. 5.9a), a combinational sleep region with at least one logic block not sharing any net with any other logic block in the sleep region (Fig. 5.9b), or a sequential sleep region that contains one or more loops (Fig. 5.9c).

For a combinational connected sleep region, as shown in Fig. 5.9a, the algorithm starts by converting the logic blocks inside the sleep region into an undirected graph. The graph in Fig. 5.10a is equivalent to the sleep region in Fig. 5.9a. Afterwards, a topological sorting of the resulting graph is used to find the relationship between all the logic blocks in the sleep region by converting the graph to a hierarchical data structure. An example of the topological sorting procedure is shown in Fig. 5.10, where A is found as the parent node, B and C are ordered in the *same level*, and D is in the last level. The linear approximation for the discharge current for the logic blocks in the sleep region is shown in Fig. 5.10e as well as the resulting sum of the discharge current vectors. It should be noted that the summation in Fig. 5.10e assumes that the logic blocks will have nonmutually exclusive discharge.

For a combinational graph with unconnected nodes, as shown in Fig. 5.11, instead of using the triangular approximation as discussed before, the discharge current is

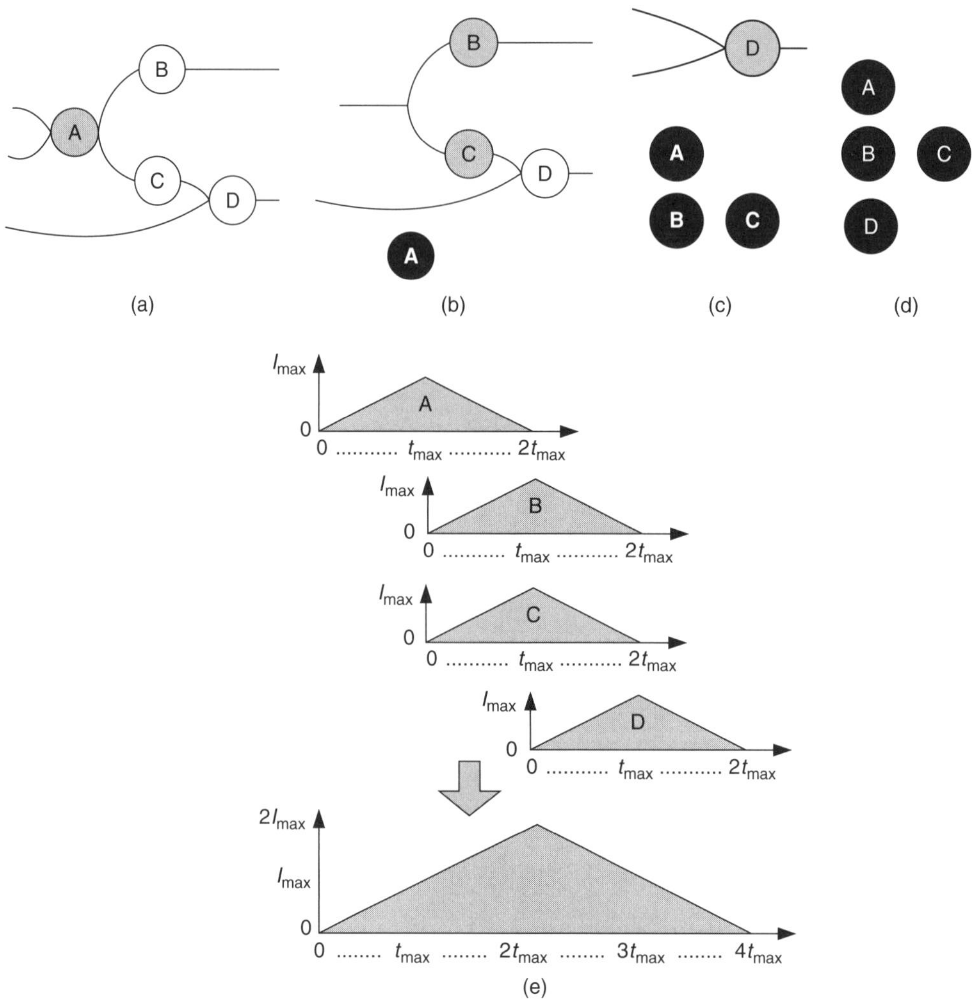

■ **FIGURE 5.10** Steps of the current feasibility check for a combinational connected sleep region. (a) A is selected to be deleted, (b) A is ordered in the first position and B and C are selected for deletion, (c) B and C are ordered in the same position, (d) Final ordering, (e) Current vectors summation.

assumed to be constant and equal to the peak value for the unconnected logic blocks because it is difficult to predict when the unconnected node is expected to discharge. The unconnected node is identified only during

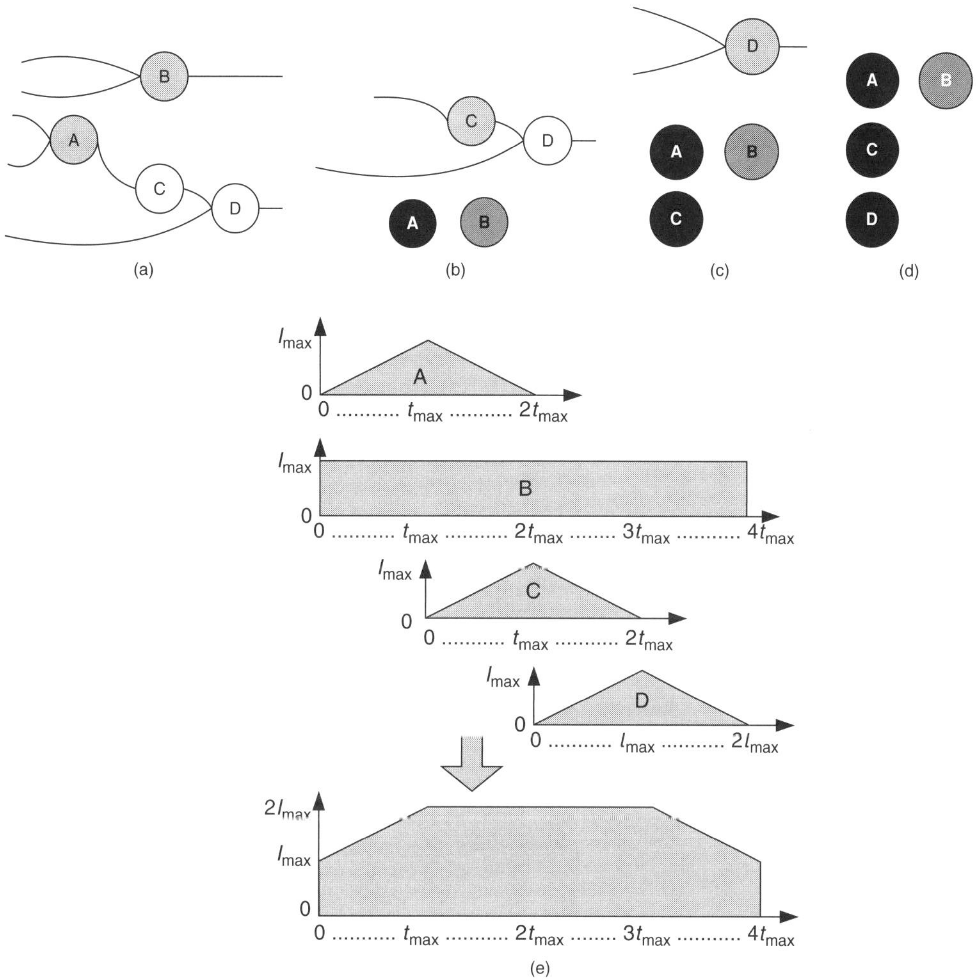

■ **FIGURE 5.11** Steps of the current feasibility check for a sequential connected sleep region. (a) Both A and B are selected for deletion, (b) A and B are ordered on the same position, with B an unconnected node and C is marked for deletion, (c) C is ordered in the next position, (d) Final ordering, (e) Current vectors summation.

the first iteration of the algorithm. Figure 5.11a represents the graphical representation of the sleep region in Fig. 5.9b. Node B is identified as an unconnected node. The algorithm then continues as the previous case to sort the rest of the graph. Afterwards, the current vector of the unconnected node B is represented as a rectangle with width equal to the sum of the widths of the other vectors and added to the rest of the currents, as shown in Fig. 5.11e.

The last case is when the graph contains one or more loops. Having a loop in the graph makes the topological ordering infeasible; hence, a loop has to be detected before starting the topological sorting algorithm. Thus, before the topological sorting phase, loop detection is used on the sleep region graph; if a loop is found, then a loop-resolving algorithm is used. It was found out that the presence of loops does not change the value of the peak current of the sleep region; it only affects the shape of the discharge current pattern. Whenever a loop is detected, it is broken at any point, and a virtual edge to represent the broken edge is kept. Afterwards, the algorithm continues in the same manner as earlier.

A pseudocode for the discharge current processing algorithm with topological sorting is listed in Algorithm 5.1. In the first step in Algorithm 5.1, the logic block under consideration is added to the cluster under consideration. Following topological sorting (Top_Sort), the logic blocks are checked according to their order in the sorting. If a block is found to be unconnected, then its current vector is treated as a rectangle with a maximum of $I_{\max}$ and starting time of 0 ($rect(I_{\max}, 0)$), as shown in Fig. 5.11e. If two blocks are sorted in the same level, as blocks B and C in Fig. 5.10d, then either their triangular discharge currents start at the same time or only one of them is considered, depending on whether they are mutually exclusive from the block on the upper level or not. Similarly for blocks from different levels, their triangular discharge current depends on whether their discharge current is mutually exclusive or not.

5.4 ACTIVITY PROFILE GENERATION

To properly explain this phase of the CAD flow, a few definitions will first be presented. An *activity profile* is a

Algorithm 5.1 Pseudocode of the logic-based discharge current processing algorithm [127]

```
for each block B to be added to activity region C do
  C = C + B
  Sort_C = Top_Sort(C)
  curr_C = 0
  for i = 0 : size(Sort_C) − 1 do
    if block_i is unconnected then
      curr_i = rect(I_max, 0)
    else
      if level_i == level_{i−1} then
        if !(block_i ⊕ block_{i−2}) then
          curr_i = trig(I_max, (i − 1) × t_max)
        end if
      else
        if (block_i ⊕ block_{i−1} ) then
          curr_i = trig(I_max, i × t_max)
        end if
        curr_C = curr_c + curr_i
      end if
    end if
  end for
  if max curr_C > I_sleep then
    C = C − B
  else
    break
  end if
end for
```

representation of the periods that a logic block is active (switching). If a group of logic blocks is expected to switch in the same time periods, then it is said that they have similar activity profiles. To maximize the power savings from the use of sleep transistors, logic blocks with similar activity profiles should be packed together and connected to one sleep transistor. The main goal of the activity generation is to identify the logic blocks that have similar activity profiles so that the packing algorithm can cluster them together. By the end of this phase, all the logic blocks in the design are given labels to divide them into several *activity regions* according to their activity profile. Logic blocks with similar activity labels have similar activity profiles. In references [126, 127], two activity profile generation algorithms are proposed: connection-based activity profile (CAP) generation and logic-based activity profile (LAP) generation, as well as a modification for the LAP algorithm reverse-LAP (R-LAP).

5.4.1 Connection-Based Activity Profile Generation Algorithm (CAP)

The main criterion used by CAP to identify the activity profiles is *connectivity*. Logic blocks that are connected to each other are expected to have similar activity profiles. The main reasoning behind this assumption is that whenever the inputs to a logic block change, its output is expected to change as well, which in turn will cause the logic blocks connected to its output to switch too. This is a pessimistic approximation of the real case as the change in the output depends on the logic implemented by the logic block.

The algorithm begins with the circuit primary inputs and greedily allocates activity regions as it traverses the circuit netlist by means of a simple depth-first graph search algorithm, resulting in a fast and computationally efficient algorithm. While traversing the circuit netlist, whenever a new logic block is reached, it is necessary to determine whether to add this logic block to the current activity region or to place it in a new activity region. There are two principal driving costs that need to be considered at each node: the size of the activity region and the attraction of a certain logic block to that activity region.

Reducing the size of the activity region provides the clustering algorithm with more flexibility to pack only those logic blocks that manifest the same activity, not those that have close activity profiles. Although this leads to a greater leakage savings, increasing the number of activity regions results in increasing the number of sleep signals used, thus causing a power-inefficient implementation, as well as complicating the control circuitry for generating these signals. Furthermore, the algorithm must be expansive while each logic block is processed. The addition of any logic block to the current activity region implies the addition of all of its fanin and fanout logic blocks, because the algorithm is connection based. Consequently, the number of fanins and fanouts of any logic block should be considered during the process, and the cost of adding the current logic block to the current activity region is expressed as

$$\text{cost}_1 = \frac{\text{currCap} + \alpha \times \text{Neighbors} - \text{maxCap}}{\text{maxCap}}, \qquad (5.8)$$

where maxCap is the predefined maximum capacity for the activity region, currCap is the current capacity of the activity region, Neighbors is the number of logic blocks *directly* connected to B and not yet placed in any activity region, and α is a weighting constant to control the quality of the final solution. The use of Neighbors provides the cost function with the capability to look around the current logic block to examine which other logic blocks are expected to be attracted to the current activity region, if the logic block under investigation is placed in it. It should be noted that the value of Neighbors can be easily evaluated during the file parsing stage without the need for a special preprocessing phase. The parameters that need to be tuned in (5.8) are maxCap and α.

The value of maxCap should be a function of the circuit size to ensure its scalability with the different circuits. In the study by Hassan et al [126], maxCap is selected as a function of the number of logic blocks on the longest path in the circuit. Reducing maxCap enables the activity generation algorithm to pack only those logic blocks that manifest the same activity, i.e., *closely* connected to each other, not those that have close activity profiles, thus resulting in a large number of activity regions, as well as sleep regions. Although this leads to more leakage savings, yet increasing the number of activity regions results in increasing the number of sleep signals used, thus complicating the control circuitry needed for generating these signals. On the other hand, a large value for maxCap will result in a large activity region with a short sleep time, thus reducing the leakage power savings resulting from the algorithm.

By running the algorithm on several benchmarks for a wide variety of values for maxCap, it was found that a value for maxCap of 1.5 times the longest path from input to output in the circuit provides the best results in terms of power savings. The average leakage power savings across the tested benchmarks is plotted in Fig. 5.12. Increasing maxCap more than 1.5 times the longest path in the circuit results in excessively large activity regions that have limited leakage power saving capability. On the other hand, decreasing maxCap increases the number of activity regions in the final design, thus resulting in a complex and power-hungry sleep-signal generation circuitry.

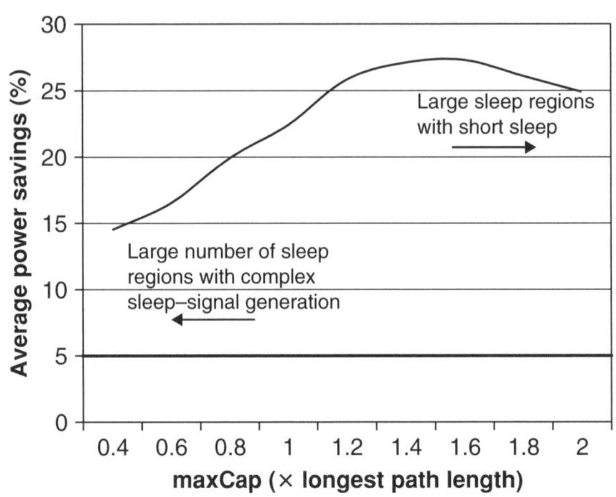

■ **FIGURE 5.12** Leakage power savings versus the maximum activity region capacity.

On the other hand, α controls the expansive ability of the algorithm. The value of α should range between 0 and 1, where a 0 value means that the algorithm considers that adding the current logic block to the cluster will not attract other logic blocks to it. A value of 1 for α means that adding the current logic block to the cluster will result in adding all the logic blocks connected to it as well. The value of α is updated adaptively according to currCap based on the following relation [126]:

$$\alpha = \begin{cases} 0.3 & \text{currCap} < 0.5 \times \text{maxCap} \\ 0.6 & 0.5 \times \text{maxCap} \leq \text{currCap} < 0.7 \times \text{maxCap} \\ 1 & 0.7 \times \text{maxCap} \leq \text{currCap} \end{cases}$$

The second cost function is the attraction between the logic block B and the current activity region C, which is expressed as

$$\text{cost}_2 = \mathit{Nets}(\text{B}) \cap \mathit{Nets}(\text{C}), \tag{5.9}$$

where $\cap$ denotes the number of nets shared between B and C. The decision of whether or not a certain logic block should be added to the current activity region resolves to a comparison between cost_1 and cost_2

$$\delta \times \text{cost}_2 - \text{cost}_1 \geq 0 \Rightarrow \text{add to the current activity region}$$
$$\delta \times \text{cost}_2 - \text{cost}_1 < 0 \Rightarrow \text{start a new activity region}$$

where δ is a normalization factor. It should be noted that cost_1 is always negative, ranging from -1 to 0, unless the activity region capacity exceeds that of the maximum capacity. On the other hand, cost_2 is a positive integer. When δ is close to 0, the activity region maximum capacity maxCap is the main limiting factor to assign activity labels, AND all activity regions will have capacity equal to maxCap. When δ is close to 1, the attraction to the activity region is the driving factor for activity labeling. Thus the activity region capacity might exceed maxCap.

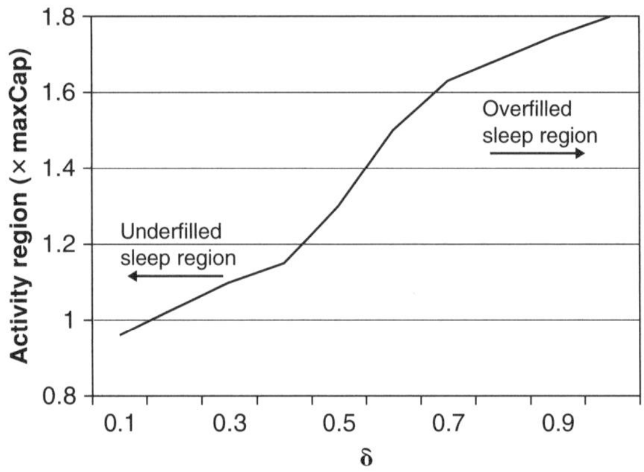

■ **FIGURE 5.13** Average activity region size across several benchmarks vs. δ.

To determine the optimum value of δ, the CAP algorithm is run several times for values of δ ranging from 0 to 1 across different benchmarks and the capacity of the activity region is recorded in each case. The average activity region size, in terms of maxCap, across all the tested benchmarks is plotted in Fig. 5.13, which shows that the average activity region size increases with δ. The value of δ used in reference [126] is 0.2, which results in an average activity region capacity of approximately 1.03 times maxCap. A value greater than that will result in larger activity regions.

Figure 5.14 depicts an example of activity generation by the modified CAP algorithm for a maximum activity region size of four. Figure 5.14 indicates that the algorithm begins with node A and then studies its child D to select the path that minimizes the total cost function, which in this case is D. Following that the child and parent of D are examined (E and B, respectively). At that point, both B and E have equal $cost_1$; hence, $cost_2$ is checked and E is selected because it has the smallest $cost_2$. The procedure continues until the algorithms start processing F, at which the activity region will be full and a new activity region will start. Hence, F and C will be in the same activity region. The pseudocode for the modified CAP algorithm is given in Algorithm 5.2.

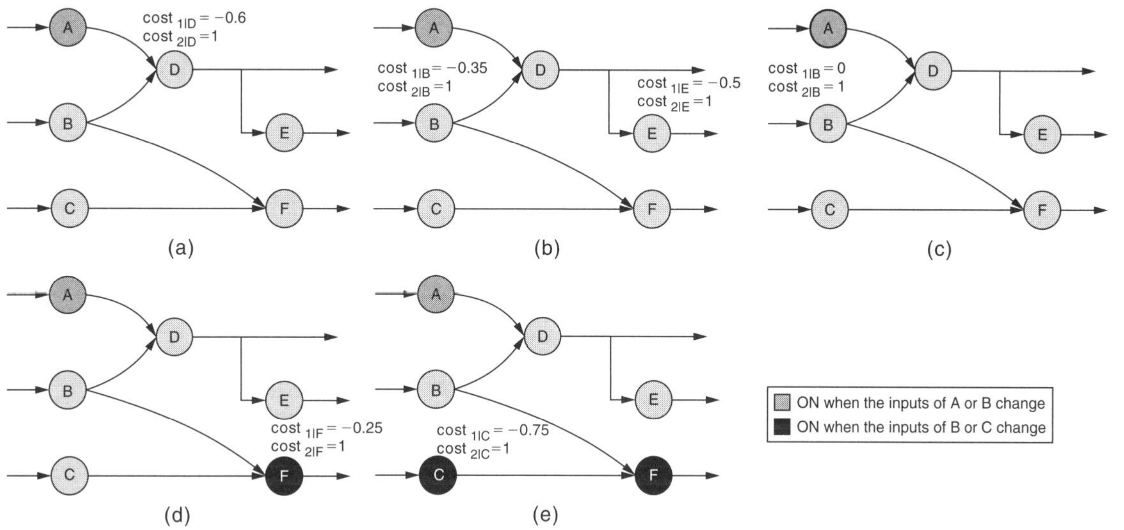

■ **FIGURE 5.14** CAP activity generation flow for maxCap $= 4$ and $\delta = 0.2$.

Algorithm 5.2 Pseudocode of the CAP algorithm [126]

Create an undirected graph from the netlist
Traverse the graph using DFS
for each node *i* **do**
  for each node *j* connected to *i* **do**
    calculate $cost_{1,j}$ and $cost_{2,j}$
    if $cost_{1,j} \leq min\_cost_1$ **then**
      $min\_node = j$
    end if
  end for
  if $\delta \times min\_cost_2 - min\_cost_1 > 0$ **then**
    add to the current activity region
  else
    start a new activity region
  end if
end for

5.4.2 LAP Generation

The LAP generation algorithm depends on representing the activities of the logic blocks as a binary sequence. The circuit topology is ignored in the LAP algorithm and instead the circuit logic function is used to find the optimum clustering that prolongs the OFF periods of each logic block. To properly explain this algorithm, several definitions and notations will be first introduced.

5.4.2.1 *Activity Vectors*

Definition 1: Activity vector

Given a net x in a circuit netlist, the *activity vector* A_x of x is defined as

$$A_x = [a_1\ a_2\ a_3\ \ldots\ a_{2^n-1}\ a_{2^n}]^T, \tag{5.10}$$

where n is the total number of inputs to the circuit, a_i is a binary variable, which is "1" if any of the outputs of the circuit depends on net x for evaluation when the inputs to the circuit are given by the ith input vector, and T represents the transpose of the vector.

In FPGAs, each logic block has only one output; thus, the activity vector of each net resolves to be the activity vector of the logic block driving that net. The circuit in Fig. 5.15 provides an example of the operation of LAP, where the logic of each block is depicted underneath the circuit. Logic blocks F and G must be ON all of the time to generate the outputs of the circuits f and g, respectively. Consequently, the activity vectors A_f and A_g for blocks F and G, respectively, are given by

$$A_f = [\ 1\quad 1\quad 1\quad 1\quad 1\quad 1\quad 1\quad 1\ ]^T, \tag{5.11}$$

$$A_g = [\ 1\quad 1\quad 1\quad 1\quad 1\quad 1\quad 1\quad 1\ ]^T. \tag{5.12}$$

On the other hand, for computing the activity vector at the inputs of block F, it is noteworthy that block D will be only used to generate the output signal f if the input

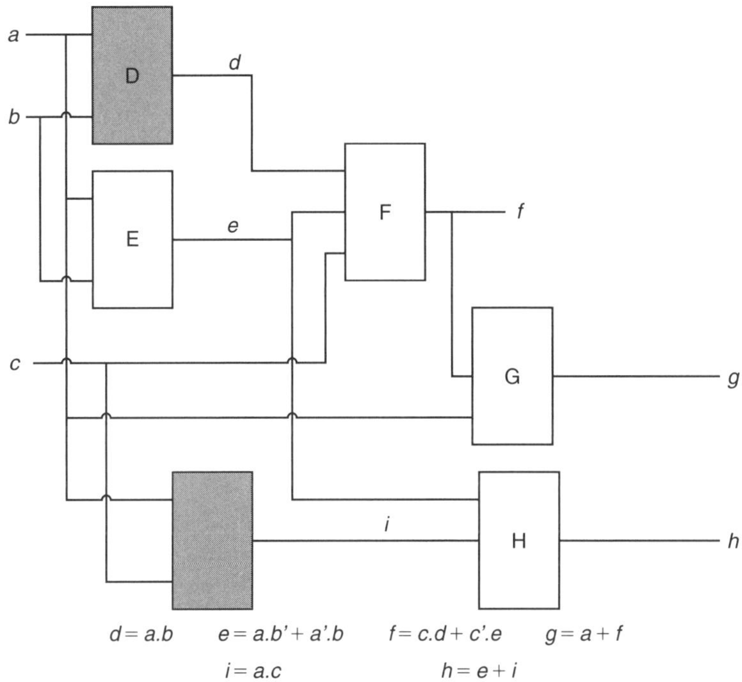

■ **FIGURE 5.15** A circuit example.

c is 1. Similarly, block E is only used when c is 0. Hence, the activity vectors for D and E, when f is evaluated, are represented by

$$A_d = [\ 0 \quad 1 \quad 0 \quad 1 \quad 0 \quad 1 \quad 0 \quad 1\ ]^T, \tag{5.13}$$

$$A_e|_f = [\ 1 \quad 0 \quad 1 \quad 0 \quad 1 \quad 0 \quad 1 \quad 0\ ]^T. \tag{5.14}$$

However, to evaluate h, E will have the following activity vector:

$$A_e|_h = [\ 1 \quad 1 \quad 1 \quad 1 \quad 1 \quad 1 \quad 1 \quad 0\ ]^T. \tag{5.15}$$

Hence, the resulting A_e is given by

$$A_e = A_e|_f + A_e|_h = [\ 1 \quad 1 \quad 1 \quad 1 \quad 1 \quad 1 \quad 1 \quad 0\ ]^T. \tag{5.16}$$

Finally, the activity vector for i is given by

$$A_i = [\ 1 \quad 1 \quad 0 \quad 0 \quad 0 \quad 0 \quad 1 \quad 1\ ]^T. \tag{5.17}$$

From this discussion, it can be deduced that if F, G, and H are active for all the input combinations, packing them together will result in improved results. Moreover, E will be active for almost all of the input combinations except for only one; thus it can also be packed with F, G, and H in the same cluster. Therefore, the cluster containing E, F, G, and H will always be ON. On the other hand, D and I have similar activity profiles for half of the input combinations; thus it will be a good strategy to group them together and turn OFF this cluster for half of the circuit operational time.

From the above discussion, it can be deduced that the complexity of the original LAP algorithm is proportional to 2^n, where n is the total number of circuit inputs. In large circuits with hundreds of inputs, this complexity renders the LAP algorithm impractical. In the next subsections, several modifications are proposed to reduce its complexity significantly.

5.4.2.2 *Hamming Distance: A Measure of the Correlation between Activity Profiles*

The relation between the activity profiles of the different logic blocks is evaluated by means of the Hamming distance between their activity vectors.

Definition 2: Hamming distance
Given two binary sequences of length n, A_n, and B_n, the Hamming distance $d_{(a,b)}$ between these two sequences is defined as

$$d_{(a,b)} = \sum_{k=0}^{n-1} |a_k - b_k|, \tag{5.18}$$

where a_k and b_k are the kth elements of A_n and B_n, respectively.

Hence, the Hamming distances between the activity vectors of the signals in Fig. 5.15 are evaluated as

$$\begin{array}{cccc} d_{(f,g)} = 0 & d_{(f,d)} = 4 & d_{(f,e)} = 1 & d_{(f,i)} = 4 \\ d_{(g,d)} = 4 & d_{(g,e)} = 1 & d_{(g,i)} = 4 & d_{(e,d)} = 5 \\ d_{(e,i)} = 5 & d_{(e,h)} = 1 & d_{(d,i)} = 4 & d_{(d,h)} = 4 \\ & d_{(i,h)} = 4 & & \end{array}$$

It can be noticed that the Hamming distance between the activity vectors of any two logic blocks is a measure of the correlation between their activity profiles. A Hamming distance close to the absolute minimum of zero indicates that the two blocks will exhibit the same activity profile, and thus, when positioned together in the same cluster, will result in maximum power savings and vice versa. This is verified by examining the values of the above Hamming distances and the results stated in the previous subsection. However, the Hamming distance between the activity vectors of two logic blocks does not take into consideration the probability of occurrence of the different input combinations, which can notably affect the quality of the results. The weighted Hamming distance is used to efficiently incorporate the various probabilities of the circuit input combinations.

Definition 3: Weighted Hamming distance

Given two binary sequences of length n, A_n, and B_n, and a weighting vector W_n, the weighted Hamming distance $dw_{(a,b)}$ between these two sequences is defined as

$$dw_{(a,b)} = \sum_{k=0}^{n-1} w_k \times |a_k - b_k|, \tag{5.19}$$

where a_k, b_k, and w_k are the kth elements of A_n, B_n, and W_n, respectively.

The use of the weighted Hamming distance is not a sufficient measure for the difference in activity between the different logic blocks. For example, if there are two logic blocks with a weighted Hamming distance of 1 between them and the net and at which they differ is an active net that continuously toggles, then the Sleep Transistor (ST) connected to them will switch frequently, increasing the dynamic power dissipation to the extent that it might override any savings in leakage power dissipation. To avoid such a condition, the transition density [65] of the net needs to be considered while calculating the Hamming distance. The transition density D is defined as the average number of transitions per unit time.

Definition 4: Transition weighted Hamming distance Given the weighted Hamming distance $dw_{(a,b)}$ between two activity vectors, A and B, and D_i the transition density of signal i, the transition weighted Hamming distance $\overline{dw}_{(a,b)}$ between these two activity vectors is defined as

$$\overline{dw}_{(a,b)} = dw_{(a,b)} \times \max\{D_A, D_B\}. \tag{5.20}$$

5.4.2.3 *The LAP Algorithm Operation*

The LAP algorithm consists of two main phases: activity vector generation and activity labeling. The activity vector generation phase exhaustively simulates the circuit by iterating all the input vectors and finding the values of all the circuit nets resulting from that input vector. Afterwards, for each input vector iteration, each signal (or block) is tested to investigate whether or not it affects the evaluation of the circuit outputs. This is performed by complementing the value of the signal under consideration and then proceeding from that point toward the circuit outputs. If the output of the logic blocks that have this net as an input will change, then this change is taken to the next circuit level. Otherwise, 0 is placed in the corresponding row of the input vector. If a loop is found, then this net is given 1 in its activity vector for that input combination. It should be noted that the number of levels checked from the net under consideration increases the computational time significantly. To limit this computational complexity, the number of levels to be checked is limited to three. After exhaustively generating all the activity vectors for all the circuit nets, the static probability of each net is calculated.

The next step is the calculation of the Hamming distance between each two logic blocks in the design. This is performed recursively through all the design elements. The transition weighted Hamming distance $\overline{dw}$ between every two logic blocks is then calculated. At this point, the activity labels can be assigned according to the weighted Hamming distance. However, this approach can result in performance deterioration as the connections between the different logic blocks is not considered. Since those logic blocks that will have a similar activity label are expected to be placed in the same sleep region, i.e., will be placed close to each other, hence, it seems that the wire length should be included in the activity labeling as well. Since at this stage the algorithm does not have any information about where each block will be placed, an approximation for the wire length is adopted. If any two logic blocks share one net, then the distance between them, l, is considered zero, e.g., E and F in Fig. 5.15. If there is one level of logic blocks in between any two logic blocks, then the distance is considered 1, and so on.

To combine the transition weighted Hamming distance and the distance between logic blocks, logic blocks are assigned activity labels by minimizing the cost function given below:

$$\min\{\overline{dw} + \delta \times l\}, \qquad (5.21)$$

where δ is a normalization constant selected to be 0.5. To avoid having activity regions with a large number of logic blocks, which will decrease the leakage savings, the size of the activity region is limited to 1.5 times the longest path from input to output in the circuit. This value was obtained by running the algorithm on several benchmarks. Assigning a constant value for activity region size, irrespective of the circuit size, results in impractical results. Increasing the activity region size more than 1.5 times the longest path in the circuit results in excessively large activity regions that are usually not fully filled up by the algorithm. On the other hand, reducing the activity region size increases the number of activity regions in the final design.

Hence, the algorithm starts to greedily assign activity labels to the logic blocks according to Eq. (5.21) until the maximum activity region size is reached. Afterwards, a new logic block is selected as a seed cell for a new activity region and the procedure is repeated. The pseudocode of the algorithm is listed in Algorithm 5.3.

5.4.2.4 *Reverse Logic-Based Activity Profile (R-LAP) Generation Algorithm*

In this section, the R-LAP generation algorithm is presented. The R-LAP algorithm is a modification of the

Algorithm 5.3 Pseudocode of the LAP algorithm [127]

```
for all the input combinations do
   for all the nets in the circuit do
      find the value of the net
   end for
   for each net in the circuit do
      toggle the value of the net
      Activity[input_vector][net] = 0
      proceed with the new value of the net
      if the value of any output changes then
         Activity[input_vector][net] = 1
      end if
   end for
end for
for each net in the circuit do
   find the static probability
   find the transition density
   find the distance to each net in the circuit
end for
```

LAP algorithm that offers a significant execution time improvement as well as more leakage power savings. R-LAP represents the logic block activity profiles using an activity vector, similar to the LAP algorithm.

R-LAP Algorithm Operation

To reduce the complexity of the LAP algorithm, Hassan et al [127] propose the R-LAP generation algorithm. In the R-LAP algorithm, instead of generating the activity vectors for the outputs of each logic block, R-LAP generates the activity vectors of the inputs to each logic block. This is performed by checking each logic block whether its ith input will contribute to the output when the other inputs are given by a certain combination. If the ith input is needed for evaluating the output, then "1" is placed in the input signal activity vector that corresponds to this input combination; otherwise, "0" is entered. Hence, the complexity of the algorithm is $O(2^{(m-1)} \times m)$, where m is the number of inputs to the logic block, which is usually around four [28].

As an example, for the circuit shown in Fig. 5.15, when b is "0," a is not needed to evaluate d. On the other hand, a is always needed to evaluate e. By taking a look at logic block F, e is not needed to evaluate f when c is 1. Similarly, d is only needed when c is 1. Furthermore, f is needed to evaluate g when c is 0. Hence, the R-LAP algorithm will evaluate the activity vectors of the different signals in Fig. 5.15 as

$$A_d|_c = [0\,1] \qquad A_e|_c = [1\,0] \qquad A_f|_c = [1\,0]$$

Hence, it can be deduced that placing F and E in the same sleep region will result in maximum power leakage savings as they both can be turned OFF when c is 1.

For each logic block, the different activity vectors for all of its inputs are generated as mentioned above. As a result, each net will have p different activity vectors associated with it, where p is the number of the net fanout. To find the Hamming distance between the logic blocks inside any sleep region, a large activity vector is generated using the smaller activity vectors of each logic block inside it, and then the Hamming distance is evaluated from the sleep region activity vector. It should be noted that the R-LAP algorithm generates the activity vectors for each logic block with respect to all the other logic blocks it is connected to. Furthermore, sleep regions are usually filled by logic blocks that share connections to reduce the total wire length and enhance the final design performance. Hence, there is no need to generate the activity vectors for each logic block with respect to all the other logic blocks in the circuit that they do not share a connection with. The pseudocode of the R-LAP algorithm is listed in Algorithm 5.4.

Algorithm 5.4 Pseudocode of the R-LAP algorithm

```
for each logic block i do
  for each input j of i do
    for each of the other input combinations k do
      Evaluate the output of block i when input j is '0'
      Evaluate the output of block i when input j is '1'
      if i|j=0 == i|j=1 then
        Aj|i[k] = 0
      else
        Aj|i[k] = 1
      end if
    end for
  end for
end for
```

5.5 ACTIVITY PACKING ALGORITHMS

Modern island-style FPGAs have a hierarchical architecture, where several logic blocks are packed together to form clusters. The packing process takes a netlist of LUTs and registers and outputs a netlist of logic clusters. The main aim of the available packing algorithms is to minimize the total area (by packing clusters to their full capacity), minimize the delay (by packing LUTs on a certain critical path together [18]), and/or maximize routability (by minimizing the number of inputs to each cluster). However, the goal of minimizing power dissipation, either dynamic or leakage power dissipation, has been rarely addressed. In references [126, 127], the activity profiles obtained earlier are incorporated into the T-VPack [18] algorithm to pack logic blocks to minimize leakage power dissipation.

5.5.1 **AT-VPack**

In reference [127], the T-VPack algorithm is modified to include activity profiles; thus the modified T-VPack is called *AT-VPack*. In AT-VPack, a set of logic blocks are selected as candidates to be added to the cluster under investigation. The selection criteria for these candidate logic blocks are (1) the combined discharge current of the logic blocks inside the cluster plus the logic block to be added does not exceed I_{sleep} and (2) the activity label of the logic block to be added is the same as that of the logic blocks inside the cluster. From the pool of candidate logic blocks, the one that maximizes Attraction and satisfies the three hard constraints of VPack is selected to be added to the cluster. If AT-VPack fails to fill out all of the spaces in the cluster, the hill-climbing approach used in the original T-VPack is invoked to start filling the vacant places while satisfying both (1) and (2).

Unlike T-VPack, AT-VPack might still be unable to fill the cluster to its maximum capacity due to the additional two constraints (1) and (2). Hence, a second hill-climbing stage is used that uses simulated annealing to swap the logic blocks in the cluster with other candidate logic blocks that have not been clustered yet and then try to fill the cluster. If the set of logic blocks currently in the cluster is given by

A and the set of logic blocks that had not been clustered is called B, the algorithm swaps block i from set A with block j from set B while satisfying constraints (1) and (2) using the following cost function:

$$\min\left\{\alpha\left[\kappa\Big(\text{Attraction}(\text{A}_i) - \text{Attraction}(\text{B}_j)\Big) + (1-\kappa)\frac{\text{number of vacant places}}{\text{total number of places}}\right]\right\}, \quad (5.22)$$

where α is a variable that represents the transition weighted Hamming distance between A and B_j ($\alpha = 1 + \overline{dw}$) and κ is a weighting constant ($0 \leq \kappa \leq 1$) that is used to give importance to either filling up the cluster with any blocks or to consider the attraction force. A small value of κ would result in a faster filling for the cluster, while a decrease in the Attraction() can be tolerated, while a large value will keep the decrease in Attraction() to a minimum and accepting partially filled clusters. By performing several experiments using AT-VPack for different FPGA benchmarks, it is found that the best value for κ is 0.5. The value chosen for α forces the algorithm to start looking at first for blocks with the same activity as the cluster before looking for blocks with other activities. Even when it does look for logic blocks with different activities, it always searches for those with close activity profiles. This ensures maximum leakage savings (clusters with blocks that have different activity profiles will be *on* for a longer period).

Moreover, the cost function in Eq. (5.22) minimizes the loss in the quality of the solution, in terms of the attraction force, by minimizing the difference between Attraction(A_i) and Attraction(B_j). Similarly, the current constraint is kept as a hard constraint throughout this hill-climbing stage. By the end of this hill-climbing stage, the cluster is full to its maximum capacity. A pseudocode for AT-VPack algorithm is listed in Algorithm 5.5. In addition, the starting temperature parameter of the simulated annealing and

Algorithm 5.5 Pseudocode for the AT-VPack algorithm

```
Perform T-VPack with 2 extra constraints:
- I_cluster ≤ I_sleep
- activity_blocks in cluster is constant
while there are empty spaces in the cluster do
  for all unclustered blocks do
    find blocks i and j with min cost (Eq. (5.22))
    add i to the cluster
    if max I_cluster > I_sleep then
      remove i
    end if
  end for
end while
```

the number of inner iterations to be performed are chosen not to be large to speed up the packing process and avoid decreasing the quality of the original solution.

The ability of AT-VPack in minimizing the number of logic blocks used in the design can be verified by finding the maximum number of unfilled clusters in each benchmark. Among all the benchmarks tested, the maximum number of unfilled clusters is four, which is less than 1% of the total number of clusters in the design.

5.5.2 **Force-Based Activity T-VPack (FAT-VPack)**

The AT-VPack algorithm suffers from long execution time because it added two hard constraints to the conventional T-VPack algorithm: (1) the combined discharge current of the logic blocks inside the cluster plus the logic block to be added does not exceed I_{sleep} and (2) the activity profile of the logic block to be added is the same as that of the logic blocks inside the cluster. As a result, this algorithm suffers from a long runtime. In reference [127], the FAT-VPack algorithm is proposed to reduce the complexity of the AT-VPack algorithm by getting rid of one of the added hard constraints used in AT-VPack. The activity profiles of the different logic blocks are added to the Attraction() function in Eq. (1.7) by including a new activity gain function. It should be noted that the constraint on the maximum discharge current of the activity region is still adopted in the FAT-VPack algorithm.

The ActivityGain is a representation of how close is the activity vector of block B to that of cluster C. In the R-LAP algorithm, the ActivityGain(B,C) of adding block B to cluster C is calculated as

$$\text{ActivityGain(B)} = \frac{2^n - \overline{dw}_{(b,c)}}{2^n}, \tag{5.23}$$

where n is the number of cluster inputs.

The total FAT-VPack gain function used in the study by Hassan et al [127] is given by

$$\text{Attraction(B)} = (1-\alpha) \times \Big[\lambda \times \text{Criticality(B)} + (1-\lambda) \times \text{SharingGain(B, C)}\Big] + \alpha \times \text{ActivityGain(B, C)} \tag{5.24}$$

where α is weighting constant ($0 \leq \alpha \leq 1$). Setting a large value for α will force the packing algorithm to pack the blocks that have the shortest Hamming distance in the same cluster—hence, same sleep region, without giving much weight to the timing information and wire length. In the following experiments, α will be set to 0.5 and the impact of its value on both the leakage savings and speed penalty will be discussed later.

5.5.3 Timing-Driven MTCMOS (T-MTCMOS) AT-VPack

In both the AT-VPack and the FAT-VPack algorithms, the total discharge current constraint was kept as a hard constraint for all the clusters. In the T-MTCMOS algorithm, the maximum discharge current is varied from one cluster to the other. From Eq. (5.6), it can be noticed that for the same $\frac{W}{L}\big|_{\text{sleep}}$ of the sleep transistor, the performance loss depends on I_{sleep}. A sleep region with a large I_{sleep} will have a larger performance loss than another one with smaller I_{sleep}, if they use equal-sized sleep transistors.

Hassan et al [128] make use of this observation to avoid incurring a large performance penalty on the critical path. Hence, the maximum performance loss along the critical path can be limited to a value smaller than that along non-critical ones, i.e., timing-driven MTCMOS (T-MTCMOS). The timing information of the logic blocks is used to vary I_{sleep} of each cluster according to its criticality using the proposed T-MTCMOS technique.

The maximum discharge current inside any cluster should not exceed the value used in Eq. (5.6) for a speed penalty of x%. The value of the discharge current can vary from one cluster to the other depending on the criticality of each cluster—hence, the speed penalty imposed on the cluster. To account for the different criticalities along the signal paths, $I_{\text{sleep}}(\text{C})$ is formulated as

$$I_{\text{sleep}}(\text{C}) = \left[1 + \delta\left(1 - \frac{\text{Criticality(C)}}{\text{Max\_Criticality}}\right)\right] \times \widehat{I}_{\text{sleep}}, \quad (5.25)$$

where $\widehat{I}_{\text{sleep}}$ is the maximum discharge current calculated for the minimum performance penalty, i.e., 3%, δ is a weighting constant, Criticality(C) is the criticality of cluster C, and Max_Criticality is the criticality of the critical path(s) of the circuit. From Eq. (5.25), it can be noticed that if the criticality of the cluster is equal to the maximum criticality of the circuit, i.e., the cluster lies on the critical path, the value of I_{sleep} will be equal to that for the minimum performance penalty, i.e., 3%; otherwise, a larger value for I_{sleep} will be used and, hence, a larger performance penalty.

The weighting factor δ is used to make sure that after adding block B to cluster C, the path does not become a critical path itself. If δ is set to a value close to 0, then all the sleep regions will have an I_{sleep} very close to that for a 3% performance penalty. On the other hand, if δ is set to 1.6, the sleep regions will have a wide variety of I_{sleep} values, hence speed penalties, with a maximum penalty of 8%. However, a large value for δ increases the possibility that the added performance penalty might cause some uncritical paths to become critical. By conducting several experiments on the value of I_{sleep}, it was discovered that by adopting an adaptive update technique for δ, shown below, depending on the criticality ratio (Criticality(C)/Max_Criticality), resulted in no new critical paths while having a wide variety for I_{sleep} values.

$$\begin{aligned} 0 < \text{Criticality(C)/Max\_Criticality} \leq 0.5 \quad & \delta = 1.6 \\ 0.5 < \text{Criticality(C)/Max\_Criticality} \leq 0.8 \quad & \delta = 0.8 \\ 0.8 < \text{Criticality(C)/Max\_Criticality} \leq 1 \quad & \delta = 0 \end{aligned}$$

5.6 POWER ESTIMATION

To evaluate the performance of the proposed algorithms, the power dissipation in the placed and routed design is compared to that of the same benchmark without sleep transistors. The power model proposed in Chapter 3, which calculates dynamic, short-circuit, and leakage power, is used to estimate the power dissipation in the design without sleep transistors. To measure the power dissipation in the design with sleep transistors, several modifications are added to the power model.

There are two standby modes for any circuit: full standby and partial standby. In the full standby state, the whole circuit is in the idle state and all the sleep transistors in the circuit should be turned off. During that period, the circuit only consumes standby leakage power. During partial standby, some parts of the circuit are in the active state and other parts are in the idle state. Hence, some of the sleep transistors are turned ON and others are OFF. Thus, the circuit will consume a combination of dynamic power and active and standby leakage power.

The total power dissipation P_t is expressed as

$$P_t = t_{on} \times P_{on} + t_{off} \times P_{idle}, \tag{5.26}$$

where t_{on} and t_{off} are the percentages of ON and OFF times of the FPGA, respectively, and P_{on} and P_{idle} are the power dissipation during the active and idle modes of operation of the FPGA, respectively. P_{on} is expressed as

$$P_{on} = [P_{dyn} + P_{sckt} + P_{leak}]_{utilized} + P_{leak}|_{unutilized}, \tag{5.27}$$

where P_{dyn}, P_{sckt}, and P_{leak} are the dynamic, short-circuit, and active leakage power dissipations, respectively, in the utilized logic blocks, while $P_{leak}|_{unutilized}$ is the standby leakage in the unutilized logic blocks.

Three different modifications were done to the power model presented in Chapter 3. (1) Leakage current is calculated only due to the sleep transistor rather than calculating the leakage through all the devices in the circuit because the sleep transistors act as a bottleneck for the subthreshold leakage current. (2) Short-circuit power dissipation is approximated as 15% of the dynamic power dissipation rather than the 10% used in the original model to account for the increased rise/fall times of the logic blocks with sleep transistors. The 15% approximation was evaluated by simulating logic blocks with and without a sleep transistor using HSpice. (3) The dynamic power consumed in the sleep transistor during switching between the ON and OFF states is calculated and added to the total power dissipation.

5.7 RESULTS AND DISCUSSION

In this section, the capability of the proposed activity profile generation algorithms presented in Section 5.4 and the packing algorithms discussed in Section 5.5 to reduce leakage power dissipation will be tested. The CAP, LAP, and R-LAP algorithms are integrated into the VPR tool together with the AT-VPack, FAT-VPack, and T-MTCMOS. In addition, the power model presented in Chapter 3 with the modifications discussed in Section 5.6 is used to estimate the power savings achieved by each combination of activity profile generation algorithm and activity packing algorithm in the final design. It should be noted that the proposed logic-based discharge current processing algorithm is used to calculate the discharge currents of the sleep regions. The proposed algorithms are tested on several FPGA benchmarks to assess their capability in minimizing both standby and active leakage power dissipation.

The decision whether to keep a utilized sleep region ON always or dynamically switching between ON and OFF depending on its activity profile is based on a balance between the leakage power savings resulting from turning it OFF during its idle periods and the dynamic power dissipated in the sleep transistor during the transition. Whenever the dynamic power dissipated in the sleep transistor during waking up or putting the cluster into sleep exceeds the leakage savings from any cluster, the cluster is kept always ON. Leakage power savings can be achieved when the cluster stays OFF for a certain period of time, $T_{\text{break even}}$. In the studies by Hassan et al [126–128], the transition density [65], the average number of transitions per cycle, is used as a measure of how long a signal stays in a certain state. Based on the transition density of each sleep signal, if the signal experiences a large number of transitions such that $T_{\text{break even}}$ is never or rarely reached, the sleep region is kept always ON; otherwise, it is dynamically turned ON and OFF depending on the activity profile.

5.7.1 Experimental Setup

In the first set of experiments, the sleep region size is set to one cluster and each cluster has four logic blocks.

This is only a starting point for the experiments and later on the optimum size of the sleep region will be evaluated. Moreover, the selected size is close to the optimum sleep region size reported in the study by Rahman et al [137].

It should be noted that the maximum allowable performance loss due to the sleep transistors in all the benchmarks is kept fixed at 5% in the AT-VPack and FAT-VPack algorithm, i.e., x in Eq. (5.6) is set to 0.05. However, in the case of the T-MTCMOS algorithm, the value of x is varied depending on the criticality of the logic blocks in the sleep region.

All the circuits tested are mapped onto the smallest square FPGA array that can accommodate them, i.e., maximum utilization percentage. The case where the design is mapped onto the minimum FPGA array is called 100% utilization. Moreover, the design is assumed to be operating without standby periods for the whole benchmark. This case is referred to as 100% ON time. Initially, the results reported are for a 90 nm CMOS process; however, toward the end of this section, the proposed leakage reduction algorithms are applied to 130, 65, and 45 nm CMOS technologies.

5.7.2 **Algorithm Comparison**

Table 5.1 lists the results of applying the different activity generation algorithms with a variety of the proposed activity packing algorithms on several FPGA benchmarks under the above-mentioned conditions. The power dissipated by each design is calculated using the modified power model discussed in Chapter 3 and the percentage savings in the total power are listed in Table 5.1. It should be noted that Table 5.1 does not iterate the results from all the possible combinations of the proposed algorithms. Only the combinations that achieve large leakage power savings are reported.

Table 5.1 Leakage Power Savings for the Different Activity Profile Packing Algorithms across Several FPGA Benchmarks

Benchmark	% of Unutilized Clusters	% Savings in Power (100% on Time)			
		CAP & AT-VPack	LAP & AT-VPack	R-LAP & FAT-VPack	R-LAP & T-MTCMOS
alu4	4.5	10.82	17.23	37.05	61.78
apex2	2.48	10.35	15.54	29.09	55.8
apex4	2.16	8.59	13.38	25.74	50.91
bigkey	2.72	9.59	14.92	30.42	56.62
clma	0.76	7.37	12.09	24.48	50.82
des	0.25	7.49	11.99	24.47	51.97
diffeq	6	11.51	17.74	36.07	57.48
dsip	4.7	8.79	14.18	30.23	50.49
elliptic	5.9	10.2	16.42	32.03	51.24
ex1010	0.26	7.73	12.09	23.07	48.49
ex5p	6.92	11.14	17.51	35.92	54.3
frisc	1.56	8.08	12.97	24.51	50.98
misex3	2.21	9.37	14.57	29.86	58.33
pdc	0.78	6.8	11.29	21.5	46.34
s298	8.32	14.28	21.06	40.94	56.75
s38417	5.6	12.46	18.21	36.34	60.19
s38584.1	1.56	8.17	12.93	25.73	51.51
seq	0	4.62	7.14	15.23	32.03
spla	3.6	8.73	12.85	27.41	50.08
tseng	8.3	13.25	19.75	39.32	56.37
Average Leakage Power Savings (%)		9.47	14.69	29.47	52.62

The power savings presented in Table 5.1 show that the combination of the R-LAP and the FAT-VPack algorithms provide more leakage power savings than the combination of CAP and LAP with AT-VPack. Furthermore, integrating the T-MTCMOS algorithm with R-LAP results in the highest power savings.

The combination of logic-based discharge current processing, R-LAP, and FAT-VPack result in higher power savings than the combination of CAP and LAP with AT-VPack because the FAT-VPack algorithm can cluster logic blocks that have close activity profiles, not necessarily the same activity profile, by finding a correlation between their activity profiles, hence, achieve more leakage savings. On the other hand, the AT-VPack algorithm can only pack the logic blocks with the same activity profiles in the same cluster; otherwise, the cluster is left ON at all times. In addition, the logic-based discharge current processing algorithm provides the packing algorithm with the flexibility to pack those logic blocks that have similar activity profiles without violating the discharge current constraint. As a result, more power savings can be achieved.

On the other hand, the T-MTCMOS algorithm achieves more power savings than FAT-VPack across all the benchmarks. The average improvement in the power savings is almost 50%. The main reason behind the increase in power savings is that in FAT-VPack the discharge current constraint is a hard constraint across all the benchmarks; thus the algorithm might fill a cluster with logic blocks that have different activity profiles and satisfy the current constraint, although there are other blocks that have closer activity profiles but violate the current constraint. On the other hand, T-MTCMOS allows the current constraint to be violated to a certain extent along noncritical paths, thus giving more freedom to the packing algorithm to pack logic blocks with close activity profiles to achieve more leakage power savings.

5.7.3 Impact of Activity Packing on Performance

As mentioned earlier, the use of sleep transistors results in a performance penalty due to the added resistance of the sleep transistor to the ground. Moreover, both FAT-VPack and T-MTCMOS do not result in the same packing as that found by the conventional T-VPack, because of the added constraints, either discharge current or activity profile, or gain functions to the optimization problem. Hence, the resulting packing might suffer from an additional speed degradation because of that reason. In the study by Hassan et al [126–128], the performance loss of the critical path is considered an indication of the performance loss for the whole design. In this experiment, the delays along the critical paths when the designs are placed and packed using the proposed packing algorithms are compared to those when the designs are packed and placed using the conventional VPR flow. Figure 5.16 plots the average speed penalties among all the benchmarks used [126–128]. Moreover, Fig. 5.16 shows the maximum and minimum speed penalties experienced in the different benchmarks for each case.

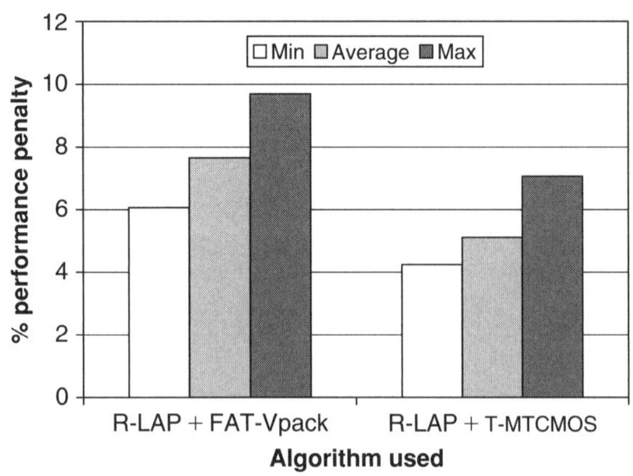

■ **FIGURE 5.16** Speed penalty experienced in the different benchmarks due to the use of sleep transistors.

From Fig. 5.16, it can be deduced that the resulting design from T-MTCMOS outperforms that of FAT-VPack in terms of timing properties. The FAT-VPack algorithm incurs a minimum of 5% delay penalty across all the paths in the design. However, T-MTCMOS increases the delay across the critical path by a minimum of 3% while making sure no other critical paths get created.

In another experiment, the maximum performance penalty allowed in T-MTCMOS is varied from 8 to 14%, while the minimum performance penalty is kept at 3% and the results for the "s298" benchmark are plotted in Fig. 5.17. It was noticed that for a sleep region of size four logic blocks, the leakage savings increased with the maximum speed penalty until a speed penalty of 10%, after which the curve almost flattens. The increase in leakage savings can be justified by the fact that increasing the maximum speed penalty allows the packing algorithm to pack logic blocks that exhibit similar activity profiles in the same

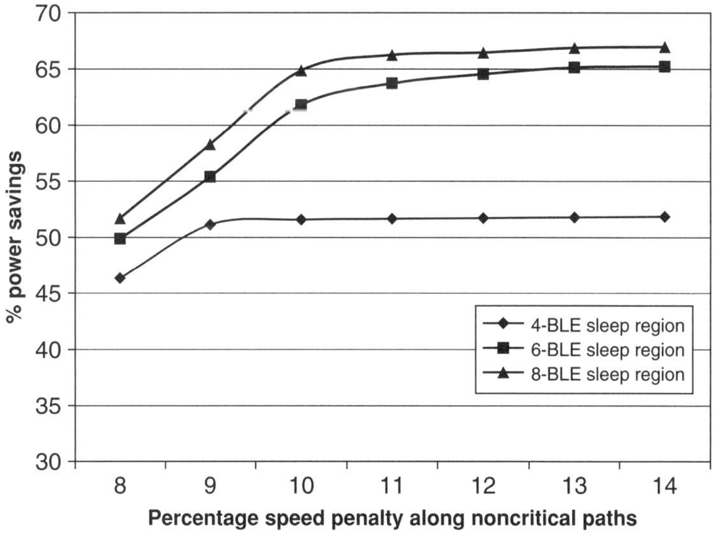

■ **FIGURE 5.17** "s298" leakage savings versus maximum speed penalty for the R-LAP and T-MTCMOS combination.

cluster regardless of their discharge current. On the other hand, as the speed penalty is increased beyond a certain limit, the packing algorithm cannot achieve more leakage savings because the sleep regions are now packed to their maximum (four logic blocks). However, as the number of logic blocks per sleep region is increased, more leakage savings can be experienced, as shown in Fig. 5.17. It should be noted that increasing the size of the sleep region beyond eight logic blocks results in an increase in the leakage savings; however, the dynamic power dissipation in the sleep transistors, which are significantly upsized, increases to cancel out most of the leakage savings.

In another experiment, the maximum speed penalty is set to 12 for the "s298" benchmark while varying the minimum speed penalty (Fig. 5.18). As the minimum performance penalty increases (by upsizing the sleep transistor), the leakage savings increases, until a certain limit after which the savings decrease because of the increase in dynamic power of the sleep transistors. It can be noticed that the breakpoint gets smaller as the size of the sleep region

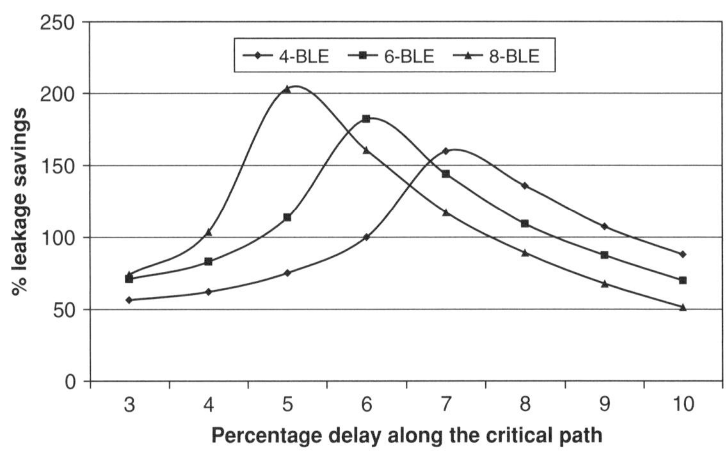

■ **FIGURE 5.18** "s298" leakage savings versus minimum speed penalty for the R-LAP and T-MTCMOS combination.

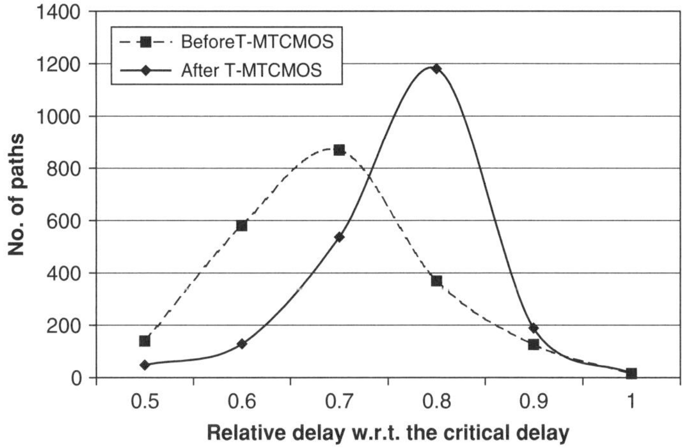

■ **FIGURE 5.19** Critical path distribution for timing-driven MTCMOS designs.

increases because larger sleep regions use large sleep transistors.

Figure 5.19 plots the relative path delay distribution in the "ex5p" benchmark with respect to the critical path delay. From Fig. 5.19, it can be deduced that the number of critical paths did not increase. In addition, the maximum circuit delay changed by only 3%. The final shape of the delay distribution can be varied by changing δ.

5.7.4 **Leakage Savings Breakdown**

In each benchmark, the leakage power savings consist of two parts: savings from permanently turning OFF all the unused clusters and savings from dynamically turning ON and OFF the used clusters in the design depending on their activity profile. By taking a look at the results for the "seq" benchmark in Table 5.1, this benchmark has no unused clusters while the power savings achieved range from 7.14 to 15.23%, depending on the combination of the activity profile generation and the packing algorithms used. This power saving is entirely from dynamically turning ON and

OFF the different used clusters in the design depending on their activity profile. On the other hand, the "s298" benchmark has the maximum percentage of unused blocks among all the benchmarks and it resulted in the maximum power savings, ranging from 21.06 to 40.94%, depending on the activity generation and the packing algorithm used.

To quantify the leakage power savings provided by the unused and used clusters of the design, the power savings from each source is recorded for each benchmark. Figure 5.20 plots the average power savings achieved by each combination of algorithms used across all the benchmarks. It should be noted that the average power savings achieved by turning OFF the unused clusters is 4.92% and is constant across all the combinations of algorithms used. This is mainly because all the combinations of algorithms provide almost the same number of clusters after packing;

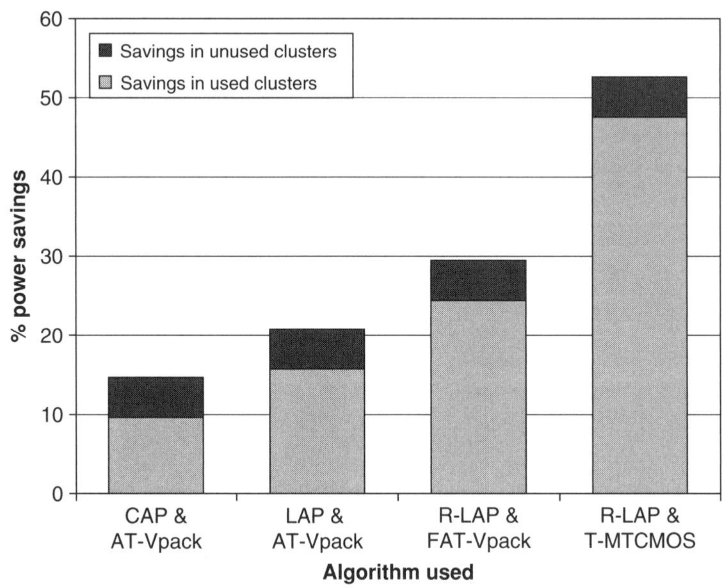

■ **FIGURE 5.20** Leakage power savings breakdown.

hence, the number of unused clusters remains the same. Figure 5.20 shows that even for the least power efficient combination of algorithms (CAP and AT-VPack), the power savings from the used clusters is almost double that from the unused clusters. The contribution of the used clusters to the total power savings increases with the algorithm efficiency.

5.7.5 **Impact of Utilization and ON Time on Leakage Savings**

In reality, the utilization percentage is less than the 100% utilization assumption used in finding the results in Table 5.1. Typically, the utilization in FPGAs ranges from 80 to 60% [129]. To investigate the impact of the utilization percentage on the total power savings by turning OFF the unused logic blocks, the benchmarks are mapped on a larger FPGA fabric and the results are plotted in Fig. 5.21 for utilization percentages of 80%

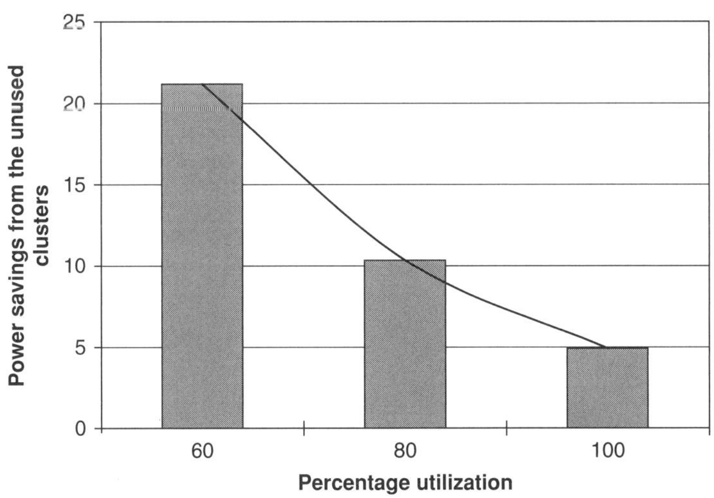

■ **FIGURE 5.21** Percentage savings in power for different FPGA fabric utilizations using the combination of R-LAP and T-MTCMOS.

and 60% using the R-LAP and T-MTCMOS combination. It can be noticed that the power savings achieved by permanently turning OFF the unused clusters increases almost exponentially with decreasing the utilization percentage.

Moreover, the 100% ON time assumption made earlier is impractically high. The average ON time of most applications is around 50 to 20% for some hand-held applications [130]. Hence, the same benchmarks are tested again using ON times of 100, 60, and 40% and the average power savings from the used clusters of the FPGA are plotted in Fig. 5.22. From Fig. 5.22, it can be noticed that reducing the operational time increases the total power savings from the proposed algorithms significantly.

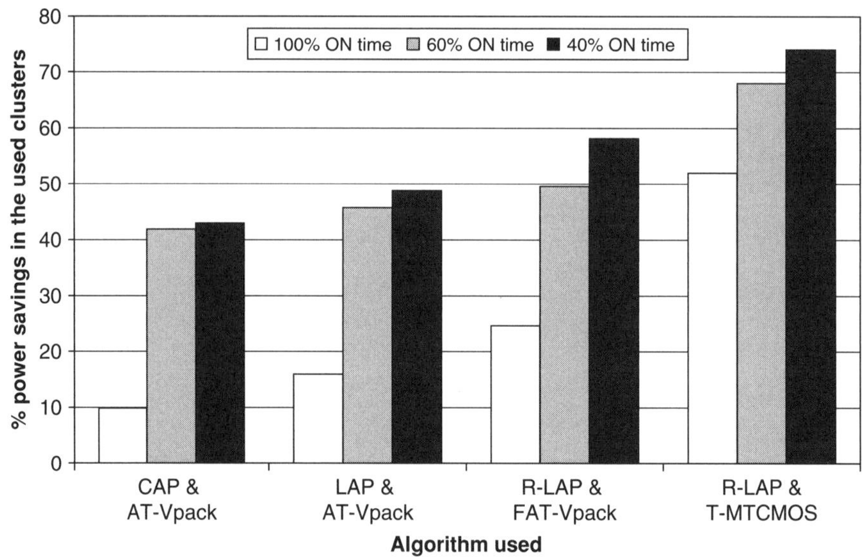

■ **FIGURE 5.22** Percentage savings in power for different utilizations and operational time using the combination of R-LAP and T-MTCMOS.

5.7.6 Impact of the Sleep Region Size

In another experiment, several sizes for the sleep region are tested for different cluster sizes. The size of the cluster is changed from three to six logic blocks and the size of the sleep region is changed from one to five clusters. The leakage savings in each of these experiments are recorded and plotted in Fig. 5.23. From Fig. 5.23, it can be noticed that for each cluster size, there is an optimum sleep region size. Moreover, leakage savings are always maximum for sleep regions of size around eight logic blocks. This proves the fact stated earlier that too large (will require a large sleep transistor, which results in large standby leakage and dynamic power dissipation in the sleep transistor) and too small (will result in partially unfilled clusters, which will increase the area and decrease the number of permanently OFF sleep regions, increasing the total leakage power) sleep regions will result in lower leakage savings.

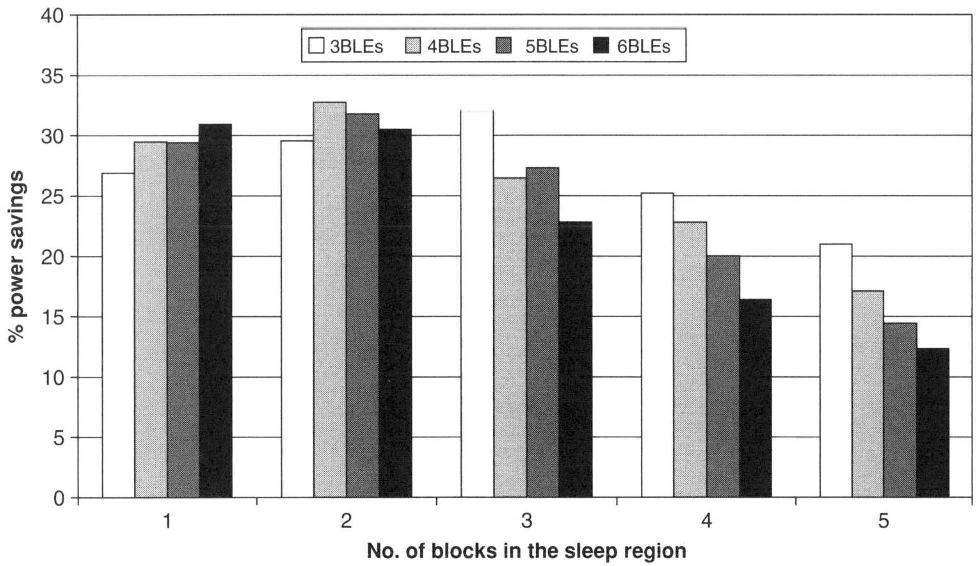

■ **FIGURE 5.23** Impact of the sleep region size on the leakage savings.

5.7.7 Scalability of the Proposed Algorithms with Technology Scaling

To investigate the scalability of the proposed algorithms, the R-LAP FAT-VPack and the R-LAP T-MTCMOS combinations are applied to several current CMOS technologies (130 nm, 90 nm) and predictive CMOS technologies (65 nm, 45 nm) [143]. The average power savings across all the benchmarks are plotted in Fig. 5.24.

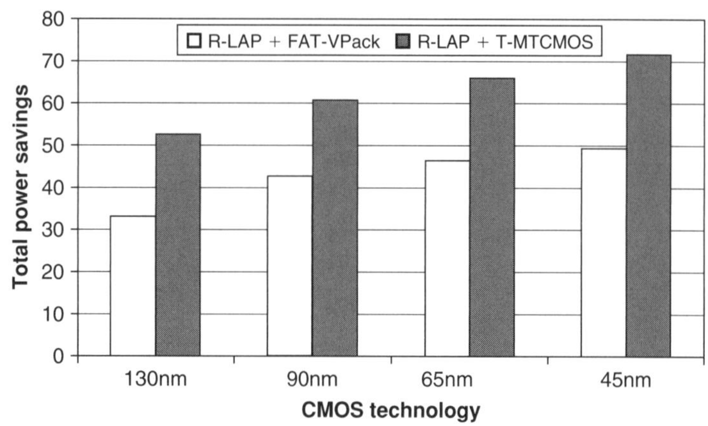

■ **FIGURE 5.24** Impact of technology scaling on power savings.

Chapter 6

Leakage Power Reduction in FPGAs Through Input Pin Reordering

Input dependency of leakage power has been witnessed in VLSI circuits in general [144] and in FPGAs in particular [44], where it was reported that four times variations in leakage power can be experienced in commercial 90 nm FPGAs.

Input signal forcing techniques have been used in the study by Anderson et al [96] to reduce the active leakage power dissipation in FPGAs. Since leakage current is heavily state dependent, by manipulating the inputs of some logic blocks, the unused parts of the FPGA can be placed in a low-leakage state. Moreover, by using the complements of the signals, the authors have managed to reduce the total leakage power of the utilized parts of the FPGA. However, the methodology [96] is based on the assumption that only one output state can result in the minimum leakage power dissipation. This is basically due to the fact that the authors only studied the power dissipation in the inverters, without trying to find a low-leakage state in the pass-transistor multiplexers. It was demonstrated that there is more than one low-leakage state that can be further exploited to achieve a bigger reduction in leakage power dissipation [145]. The technique proposed by Anderson et al [96] focuses on leakage power minimization only in the inverters and the buffers of the FPGA without considering leakage power minimization in the other parts of the FPGA, including the pass-transistor multiplexers.

In FPGAs, input signal forcing is a substantial leakage power reduction technique since FPGAs depend on pass-transistor logic in their design, where power dissipation is strongly state dependent. In this chapter, a complete new methodology, based on input pin reordering, is developed to reduce the total leakage power dissipation in all components of FPGAs, unlike the study by Anderson et al [96] that focuses only on the inverter, without incurring any area or performance penalties in the final design. In the methodology proposed in [145], the logic and routing resources are handled differently to achieve maximum leakage savings. Moreover, a modified version is implemented to improve the performance along the critical path and still achieves significant leakage power savings in the design. Moreover, the impact of technology scaling on the lowest leakage states was investigated in the study by Hassan et al [145].

This chapter is organized as follows: the state dependency of leakage power in FPGAs is discussed in Section 6.1. The input pin reordering algorithm is introduced in Section 6.2. Finally, the results of applying the pin reordering algorithm on the generic FPGA architecture are discussed in Section 6.3.

6.1 LEAKAGE POWER AND INPUT STATE DEPENDENCY IN FPGAs

Leakage current in nanometer CMOS technologies has two main components: subthreshold and gate leakage currents. Subthreshold leakage current flows from the drain to the source of an OFF CMOS device. On the other hand, gate leakage current flows through the gate of the device to or from one or both the diffusion terminals. Gate leakage has both ON and OFF components, with its OFF component almost ignorable relative to the ON part [43]. Both components of leakage power exhibit strong dependency on the state of the input signals as discussed in this section.

6.1.1 Subthreshold Leakage Current

The subthreshold leakage current I_{sub} is defined as the current that flows between the drain and the source of an MOS device when V_{GS} is less than V_{TH}. I_{sub} is formulated as

$$I_{\text{sub}} = \mu_{\text{o}} C_{\text{ox}} \frac{W}{L} (m-1) \times \upsilon_{\text{T}}^2 \times e^{(V_{\text{GS}}-V_{\text{TH}})/m\upsilon_{\text{T}}} \times (1 - e^{-V_{\text{DS}}/\upsilon_{\text{T}}}), \tag{6.1}$$

where μ_{o} is the device mobility, C_{ox} is the oxide capacitance, W and L are the device dimensions, υ_{T} is the thermal voltage (kT/q), and m is the subthreshold swing coefficient, which is given by

$$m = 1 + \frac{3t_{\text{ox}}}{W_{\text{dm}}}, \tag{6.2}$$

where t_{ox} is the oxide thickness and W_{dm} is the maximum depletion layer width. The contribution of the subthreshold leakage current to the total power dissipation increases with technology scaling due to the continuous reduction in V_{TH} to improve the device performance.

The input state dependency on the subthreshold leakage current can be readily seen in Eq. (6.1) in the dependence of I_{sub} on V_{DS} and V_{GS}. Two dominant factors affect the input dependency of subthreshold leakage current: drain-induced barrier lowering (DIBL) and body effect. Subthreshold leakage current is also a strong function of the temperature, increasing significantly with increasing chip temperature.

6.1.1.1 ***DIBL***

In nanometer CMOS devices with short channels, the drain-source potential has a strong impact on the band bending over a significant part of the CMOS device. As a result, the threshold voltage of the CMOS devices becomes a function of the drain-source voltage. Applying a large drain-to-source voltage to the CMOS device results in decreasing the threshold voltage, hence, increasing the subthreshold current. Figure 6.1 plots the change in V_{TH} and the subthreshold leakage current I_{sub} of a minimum size 90 nm NMOS device with the change in V_{DS} from 0 to 1.2 V. It should be noted that this 90 nm CMOS process has a supply voltage of 1.2 V; thus the change in the drain-to-source voltage plotted in Fig. 6.1 can be readily experienced during operation. Figure 6.1 shows that for two equal-sized transistors, their V_{TH} can differ by almost 25% and their leakage current can vary by 4.5 times, due to the DIBL effect, because of a difference in V_{DS} equal to the supply voltage.

Pass-transistor multiplexers used in FPGAs can experience four different values of V_{DS}, as shown in Fig. 6.2. The transistors in the first and last stages of the multiplexer are the

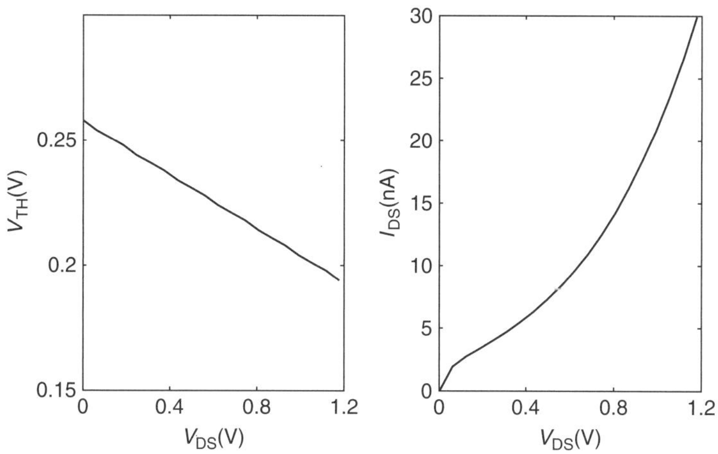

■ **FIGURE 6.1** DIBL effect in a 90-nm CMOS process.

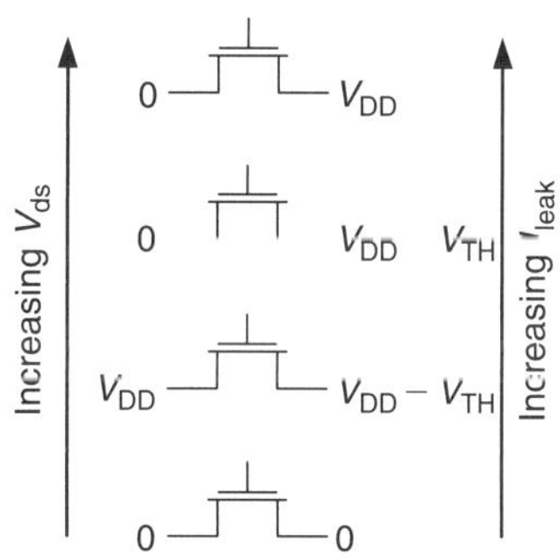

■ **FIGURE 6.2** DIBL impact on subthreshold leakage in FPGA pass-transistor devices.

only ones that can experience the worst case V_{DS} of V_{DD}. The middle stages can experience a maximum of $V_{DD} - V_{TH}$ because of the weak "1" passed by the NMOS pass transistors. From Fig. 6.2, it can be deduced that the maximum leakage current occurs when the signals at both diffusion terminals of a multiplexer transistor are different, i.e., to have the largest value of V_{DS}.

From the above discussion, it can be concluded that a means of reducing subthreshold leakage in pass-transistor multiplexers is to ensure that the majority of the pass transistors experience the smallest V_{DS}. It should be noted that if the first stage of the multiplexer is designed to have the smallest V_{DS}, the total multiplexer subthreshold current will be limited significantly, since the total subthreshold current has to flow through them.

6.1.1.2 *Body Effect*

The impact of the body to source voltage V_{BS} on V_{TH} has been witnessed in CMOS devices for several technology generations. The effect of body bias is formulated as

$$V_{TH} = V_{TH0} + \gamma\left(\sqrt{|\Phi_S| - V_{BS}} - \sqrt{|\Phi_S|}\right), \tag{6.3}$$

where V_{TH0} is the ideal V_{TH} at zero V_{BS}, γ is the body bias coefficient, and Φ_S is the surface potential. Having a negative V_{BS} would result in increasing the threshold voltage, which in turn will reduce the subthreshold leakage current.

It should be noted that CMOS devices in pass-transistor multiplexers will never experience a positive V_{BS}, since the body of the pass-transistors is always connected to GND. Pass transistors with logic "0" at one or both of the diffusion terminals will not experience body effect as V_{BS} would be zero. However, those devices with logic 1 at both their diffusion terminals will experience subthreshold leakage current reduction due to body effect because their $|V_{BS}|$ would be maximum, either V_{DD} or $V_{DD} - V_{TH}$.

6.1.2 Gate Leakage

Gate leakage exists in both the ON and OFF states of the CMOS devices [43]. However, the OFF component of the gate leakage is ignorable with respect to the ON component. The value of gate leakage is a strong function of both V_{GS} and V_{DS}. Large values of $|V_{GS}|$ and small values

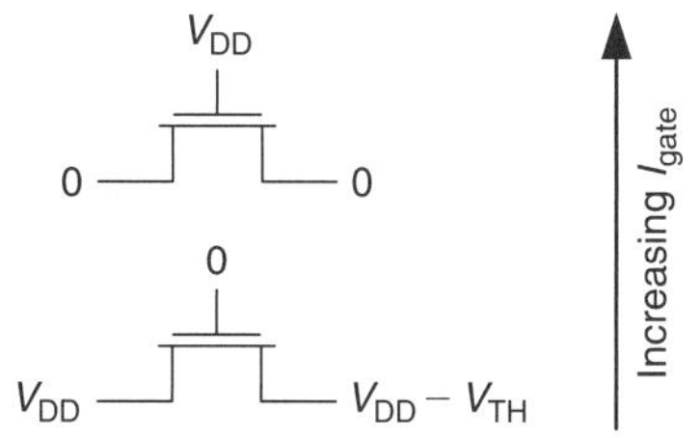

■ **FIGURE 6.3** Gate leakage dominant states in FPGA pass-transistor devices.

of V_{DS} generate a large gate leakage current. Figure 6.3 shows the two dominant gate leakage current configurations and how they depend on the input state. The gate leakage resulting from the other input configurations is much smaller than these two configurations and can be safely assumed zero, at least for the 90 nm CMOS process used in the study by Hassan et al [145]. It should be noted that the gate leakage current is not a function of the temperature and thus stays constant with the change in the chip temperature.

6.1.3 Low-Leakage States in Pass-Transistor Multiplexers

From the above discussion, it can be concluded that there is one or more input states where the leakage power will be minimum in pass-transistor multiplexers. The input state depends on the relative magnitude of the subthreshold leakage current to that of the gate leakage current. In the study by Kumar and Anis [87], it was reported that the gate leakage power dissipation is less than 1/20 of the subthreshold leakage power dissipation in 90 nm FPGAs at room temperature. In the experiments done by Hassan et al [145], the maximum gate leakage current in a 90 nm minimum-sized device is on the order of 300 pA, which is much less than the smallest subthreshold leakage current measured, which is on the order of 1 nA. Moreover, the contribution of the gate leakage power decreases with the increase in temperature due to the strong dependence of subthreshold leakage power on the temperature. Consequently, the most dominant leakage states are considered to be those of the subthreshold leakage current [145]. Figure 6.16 shows the input states that result in the lowest and highest leakage current that can be experienced in FPGA pass transistors.

The configuration labeled (1) in Fig. 6.4a results in the lowest leakage current. The highest leakage state is the one labeled (1) in Fig. 6.4b, which experiences the highest V_{DS}, hence, the maximum DIBL effect and no body effect. By looking at the low-leakage states shown in Fig. 6.4b, it can be deduced that the lowest leakage states occur if every pair of pass transistors in the multiplexer have inputs with similar value, as shown in Fig. 6.5a. The highest leakage states occur whenever the inputs to the multiplexer pair are different, as shown in Fig. 6.5b.

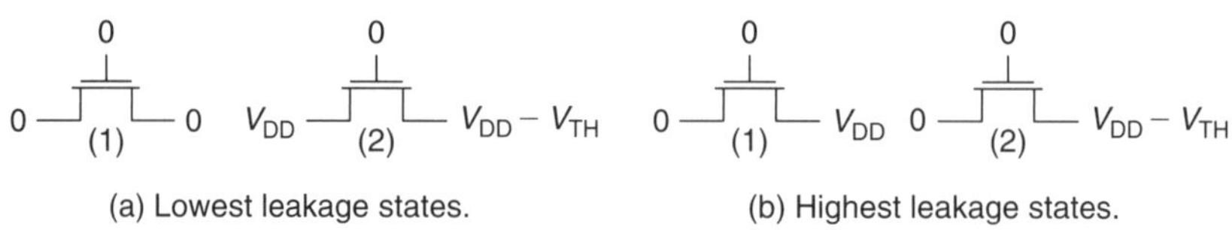

■ **FIGURE 6.4** Total leakage dominant states in FPGA pass-transistor devices.

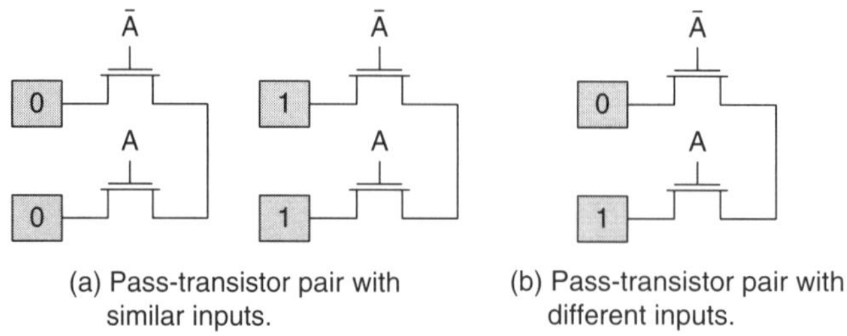

■ **FIGURE 6.5** Inputs to pass-transistors pairs.

6.1.4 Leakage Power in Inverters/Buffers

In this experiment, a minimum-sized inverter is designed to have equal rise and fall times to minimize short-circuit power dissipation. This is achieved by increasing the width of the PMOS device while keeping the NMOS device to minimum width to balance the difference in mobility of the two devices. The inverter is then simulated using HSpice and the total leakage power is recorded in both cases when the output of the inverter is 1 and 0. The values of the measured leakage current are recorded in Table 6.1. As seen in Table 6.1, even if the PMOS and NMOS devices of the inverter are designed to have equal driving capabilities, the NMOS still leaks more than the PMOS device. The ratio between the NMOS total leakage current to that of the PMOS is almost two times. This is mainly due to the inverse narrow width effect experienced by trench-isolated CMOS devices. PMOS devices in trench-isolated devices experience an increase in V_{TH} with the initial increase in the width of the device. Afterwards, the leakage current starts increasing with the device width. As a result, PMOS devices in the inverters sink a smaller subthreshold leakage current than the NMOS devices because of their larger width.

This is an interesting phenomenon as it can be used to further reduce leakage inside LUT pass-transistor multiplexers. Multiplexers need inverters to generate the complement of the control signals, which are generated inside the LUT using minimum-sized inverters similar to the one simulated above, as shown in Fig. 6.6. If the input control

Table 6.1 Leakage Current in a Minimum-Sized Inverter

Inverter Output	Total Leakage Current (nA)
0	17.03
1	31.12

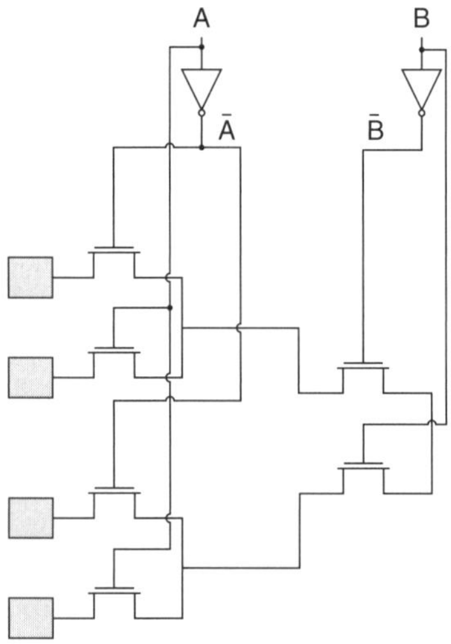

■ **FIGURE 6.6** Gate leakage dominant states in FPGA pass-transistor devices.

signal are all zeros, A and B in Fig. 6.6, then all the inverters would have a high output, thus sinking the largest leakage current. Hence, it is more leakage efficient to avoid having the most probable input state being all zeros, where all the inverters would generate the highest leakage current.

6.2 THE INPUT PIN REORDERING ALGORITHM

The pin reordering algorithm for leakage power reduction proposed in [145] uses the conclusions developed in Section 6.1 to reduce total leakage power in the pass-transistor multiplexers. The algorithm consists of two phases: the first one targets leakage reduction in the FPGA logic blocks, logic pin reordering (LPR), and the second phase targets the routing switches, routing switch pin reordering (RPR). The LPR stage is performed right after the synthesis and before the packing stage, whereas RPR is performed after the routing stage, as shown in Fig. 6.7. Again, the CAD flow used in the study by Hassan et al [145]

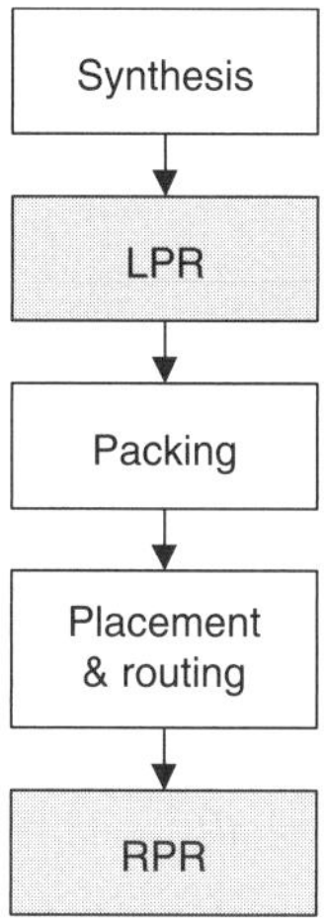

■ **FIGURE 6.7** VPR CAD flow with the pin reordering algorithm [145].

is based on the VPR CAD flow [28], where the packing is performed using T-VPack and placement and routing are performed using the VPR CAD tool.

6.2.1 LPR Algorithm

The LPR algorithm reorders the input pins in such a way to have the maximum number of signals with similar polarities at the inputs of any multiplexer pair, as shown in Fig. 6.5a. The algorithm also avoids having the input configuration with the highest probability to be the one with all zeros to further minimize the leakage power in the LUT, as explained in Section 6.2. Furthermore, the LPR algorithm handles logic blocks differently according to the number of inputs of each logic block. The LPR algorithm is divided into four separate phases as discussed in the next subsections.

6.2.1.1 *Input Pin Padding*

In most FPGA CAD tools, whenever a logic block has inputs less than the maximum number of allowable inputs to a logic block, the unused inputs are either left floating or connected to V_{DD} or GND. The choice whether to connect the unused input pin to either rail does not follow a certain reasoning, but rather it is an architecture choice. A padding methodology is used in [145] to make use of the fact that multiplexer pairs with similar inputs at both diffusion ends sink less leakage current than those with different inputs [145], as shown in Fig. 6.5.

If a logic block has inputs less than the maximum number of allowable inputs, the extra inputs are padded in such a way to create the largest number of low-leakage multiplexer pairs. As an example, assume that the maximum number of inputs for every logic block is three, and the logic block shown in Fig. 6.8a has only two inputs, A and B. The LPR algorithm then pads the extra input C to the circuit, as shown in Fig. 6.8b. C is a fixed signal that can be set to either 1 or 0. It should be noted that all modern FPGAs have the ability to generate a constant signal from within the logic block with no need to use any extra resources. As a result, the circuit with the padded inputs will have all of its first-level multiplexer pairs with identical signals at both inputs, as shown in Fig. 6.8b, and hence, maximum leakage reduction can be achieved in this case. It should be noted that the inputs padded into the circuit always go into the least significant bits of the multiplexer.

6.2.1.2 *Input Pin Swapping*

The second phase of the LPR algorithm is involved with the swapping of the input pins to have the maximum number of multiplexer pairs with similar signals at their inputs. Assume that the inputs to a 4-input logic block are $A_0A_1A_2A_3$. The algorithm picks the first input signal from the synthesized circuit A_0 and looks at the outputs of the logic function implemented when $A_1A_2A_3$ are given by "000," while A_0 is both "0" and "1." If the two outputs are equal, the counter for A_0 is incremented by 1. Afterwards, the algorithm looks at the outputs of the function

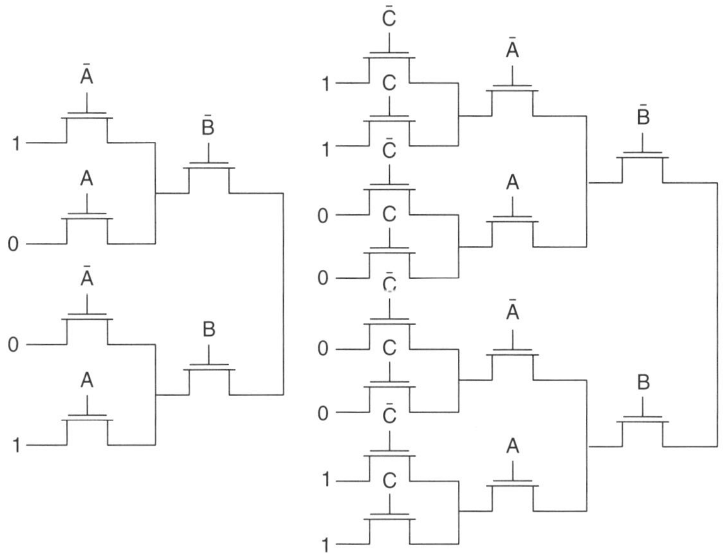

(a) Logic block with inputs less than the maximum.

(b) Resulting circuit after input padding.

■ **FIGURE 6.8** Input padding for logic blocks with inputs less than the maximum.

when $A_1A_2A_3$ are given by "001," and so on until all the 2^3 different combinations are considered. The same procedure is repeated for the other three inputs A_1, A_2, and A_3. The input with the highest count of equal signals is selected as the least significant input pin. The computational complexity of this phase is $O(m \times 2^{m-1})$, where m is the maximum number of inputs to the logic block. A conventional value of m is four [28].

As an example of the algorithm operation, consider the 2-input logic block shown in Fig. 6.9a. The count of signals with similar input multiplexers would be zero for A and two for B. Hence, the algorithm would move B to be the least significant bit of the multiplexer instead of A. The resulting logic block with configuration SRAM bits is shown in Fig. 6.9b, where it can be seen that now the first-stage multiplexers have signals with similar polarity at their inputs.

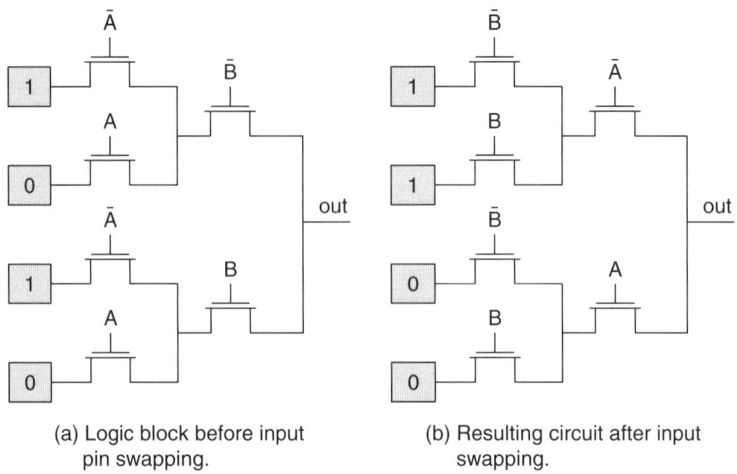

■ **FIGURE 6.9** Input pin swapping for logic blocks to minimize leakage power dissipation.

After identifying the least significant input, the algorithm tries to find the order of the remaining inputs using the same methodology. However, in finding the least significant input pin, the inputs to the multiplexer are known since they are the contents of the configuration SRAM cells. In the following multiplexer stages, static probability is used to find the most probable value expected at their inputs. If the static probability of the control input to the least significant multiplexer is higher than 0.5, the algorithm assumes that all the values connected to the transistor controlled by the least significant pin will pass to the second stage. On the other hand, if the static probability is less than 0.5, the inputs controlled by the complement of the least significant input pin are assumed to go through. The same procedure used to select the least significant pin is used to order the remaining input pins.

It should be noted that most of the leakage savings are achieved from the selection of the least significant input pin. This is mainly because the leakage current in the multiplexer will be limited to the smallest leakage current on the path. However, rearranging the pins of the latter stages adds an extra amount of leakage savings by adding more resistance in the leakage current path. It is worth mentioning that this phase of LPR does not add any physical overhead to the design since it merely rearranges the input pins and the configuration SRAM contents.

6.2.1.3 *Most Probable States*

As presented in Section 6.1.4, CMOS inverters dissipate almost two times more leakage power when their output is 1. Since the multiplexers use several inverters to generate the needed input signals to control the multiplexer, then it might be wise to avoid having the most probable input being given by all zeros. Such an input will generate the highest leakage power dissipation in the inverters inside the multiplexer. As a matter of fact, it is more desirable to have the most probable input being all ones.

To make use of this property, the LPR algorithm looks at the static probabilities of the different input combinations to each multiplexer. If the highest one contains a large number of zeros, i.e., more than half the maximum number of inputs, the algorithm tries to avoid that by inverting one or more of the input signals. This can be easily done in FPGAs as it only implies changing the contents of the SRAM cells in the logic blocks that are connected to this signal.

The LPR algorithm looks at the set of inputs that need to be inverted and tries to invert only a small number of them that is needed to counterbalance the effect of the large number of zero inputs. As an example, consider a 4-input logic block that has the most probable input being "1000." Then the algorithm would need to toggle two of the three least significant inputs to counterbalance the effect of the majority of zeros in the most probable inputs. The toggling is performed by toggling the contents of the configuration SRAM cells in the logic block that generates these signals. The choice of which signals to toggle is based on the probabilities of the next most probable inputs. If a certain signal appears 0 more than once in the first five most probable inputs states, it has a higher probability of being selected by LPR to be toggled. If the five most probable input states have static probabilities given by $P_{1,2,3,4,5,}$, the value of the input j in input state i is given by $I_{j,i}$. The inputs selected for toggling are those that maximize

$$\max_{\forall j} \sum_{i=0}^{5} I_{j,i} \times P_i \,. \tag{6.4}$$

6.2.1.4 *Unutilized Logic Resources*

The unutilized logic resources should always be placed in a low-leakage state to avoid wasting leakage current. From the discussion presented in Section 6.1 and summarized in Fig. 6.4b and Table 6.1, the lowest leakage state occurs when all the SRAM cells store 0. In addition, the input to the inverters that generate the control signals inside the logic blocks should be connected to V_{DD}. Since FPGAs have the flexibility to connect any input inside the logic blocks to either of the supply rails, then this phase of LPR guarantees placing the unutilized logic resources in a low-leakage mode without incurring any physical costs.

6.2.2 **Routing Switch Pin Reordering (RPR) Algorithm**

The routing architecture assumed by Hassan et al [145] is the disjoint architecture with a flexibility of three. The RPR algorithm is composed of three phases.

6.2.2.1 *Input Pin Padding*

The input padding phase of the RPR algorithm is similar to that of the LPR algorithm. If one of the routing multiplexers has one of its inputs left floating, that input is connected to a constant signal in such a way to reduce the leakage power in that multiplexer, according to the guidelines in Section 6.1. Similar to the logic phase, these multiplexers have the flexibility to generate constant signals from within, thus reducing the need for extra hardware.

However, the inputs to the drains of the pass transistors of the routing switches are not known beforehand, unlike the logic blocks which are dictated by the contents of the configuration SRAM cells. The work by Hassan et al [145] depends on the static input probabilities of the routed signals to estimate the most probable value of the signal and based on that pad the vacant input signals accordingly to minimize the leakage power dissipation.

6.2.2.2 *Most Probable States*

Some of the routing resources used in FPGAs are buffered routing switches, where the output of the multiplexer is connected to a buffer to transmit the signal for a long distance across the FPGA fabric. However, these buffered switches are prone to the input dependency of the leakage power dissipation in the buffers, especially since they are designed with large dimensions. This observation is the core of the earlier work [96].

The leakage power dissipation in the buffers can be minimized by avoiding the state with the highest leakage. Inside the buffer, both inverters dissipate leakage power dissipation; however, the second inverter dissipates more leakage power because it is designed with a larger size. Consequently, the desired low-leakage state is that of the second inverter. According to that, the low-leakage state is when the input to the buffer is 1. Since the input to the buffer is not previously known, static probability is used to estimate the most probable input to all the buffered routing switches. If the most probable input is not 1, then the input is inverted simply by inverting the values of the logic blocks connected to that net. It is noteworthy that this phase of the RPR algorithm is applied first since it might affect the actions taken by input padding phase. The algorithm used in the study by Hassan et al [145] is similar to the one presented in the study by Anderson et al [96].

6.2.2.3 *Unutilized Routing Resources*

Similar to the LPR algorithm, the RPR is applied to place all the unutilized routing resources in a low-leakage state. This is performed by generating constant signals within the unused routing multiplexers to manipulate them into the lowest leakage state.

6.3 EXPERIMENTAL RESULTS

The leakage power reduction pin reordering algorithm proposed in [145] is tested on a 90 nm CMOS process using several FPGA benchmarks [28]. The algorithm takes as an input a readily synthesized circuit and then rearranges the inputs to each LUT to result in the minimum leakage power dissipation. The benchmark circuits are synthesized using the SIS sequential circuit synthesis tool [146]. Both the LPR and RPR algorithms are integrated into the VPR CAD flow [28] according to the flowchart shown in Fig. 6.7. The leakage power modeling is performed using the power modeling approach proposed in [90] to take the state dependency of leakage power into consideration.

The pin reordering algorithm is applied to several FPGA benchmarks and the percentage leakage savings are listed in Table 6.2 for a 4-input LUT compared to the control case. Moreover, Table 6.2 also lists the total leakage savings achieved [96]. The total average leakage savings is around 50% across all the benchmarks tested, while that achieved by Anderson et al [96] is around 24.72%. It can be easily shown that the pin reordering algorithm outperforms that of the reference [96] in terms of leakage savings. In addition, Table 6.2 shows how the leakage savings due to the LPR algorithm vary with the number of inputs to the logic blocks. The larger the number of logic blocks with inputs less than four, the higher the leakage savings due to LPR due to its padding phase.

Table 6.2 Leakage Savings by the Pin Reordering Algorithm across Several FPGA Benchmarks [145]

Benchmark	Percentage of Logic Blocks with			Avg. Leakage Savings in Logic (%)	Avg. Leakage Savings in Routing (%)	Tot. Leakage Savings (%)	Avg. Leakage Savings (%) [96]
	Two Inputs	Three Inputs	Four Inputs				
alu4	0.07	0.29	0.62	60.92	40.76	50.3	20.7
apex4	0.01	0.42	0.55	57.34	41.35	48.9	26.5
des	0.05	20	0.74	57.28	45.74	50.96	29.99
dsip	0	0	0.98	20.13	45.72	33.16	20.86
frisc	0.07	0.27	0.64	68.34	42.18	54.61	26.16
pdc	0.01	0.21	0.76	52.07	42.28	47.18	24.5
s298	0.08	0.22	0.68	58.84	42.27	50.12	26.85
s38584.1	0.25	0.18	0.53	79.48	43.79	61.49	16.48
spla	0.01	0.24	0.74	53.78	42.8	48.02	21.2
apex2	0.06	0.31	0.62	56.68	41.51	48.78	30.36
bigkey	0.2	0	0.79	64.08	43.43	53.28	28.69
clma	0.06	0.24	0.69	49.91	42.6	46	17.79
diffeq	0.09	0.29	0.61	67.87	41.32	53.8	27.02
elliptic	0.12	0.28	0.59	60.01	44.07	51.27	27.19
ex5p	0.04	0.21	0.74	40.80	41.27	41.04	36.7
misex3	0.04	0.35	0.59	66.08	39.29	52.2	18.7
s38417	0.04	0.41	0.52	57.10	43.74	50.27	18.83
seq	0.07	0.33	0.59	68.44	41.72	54.2	24.7
tseng	0.12	0.27	0.6	70.58	43.47	56.05	19.18
ex1010	0.04	0.42	53	69.84	41.09	54.09	31.9
			Average	58.98	42.38	50.29	24.72

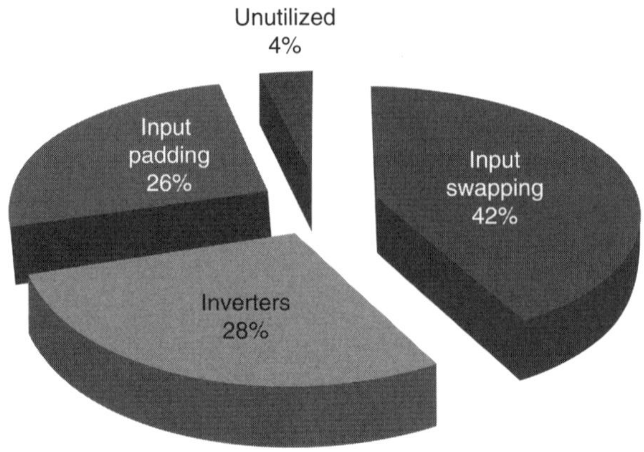

■ **FIGURE 6.10** Leakage savings breakdown in logic blocks.

Figure 6.10 shows a breakdown of the leakage savings due to the LPR phase. It can be shown that the maximum savings are generated by the input swapping phase. Figure 6.10 shows that very small leakage savings originate from placing the unutilized logic into a low-leakage state because the benchmarks tested were mapped into the smallest FPGA square array that can hold them, thus resulting in the absolute minimum unutilized logic resources. It should be noted that if the benchmarks were mapped into more practical FPGA sizes, the percentage leakage savings in the unutilized resources would increase notably.

The leakage power savings achieved using RPR breakdown is shown in Fig. 6.11. It can be noticed that the average savings in the unutilized routing resources is larger than that resulting for the logic resource. This is mainly because the percentage of unutilized routing resources is usually larger than that of the logic resources. Another observation is that the percentage leakage savings in the inverters is larger in the routing resources, which can be justified by the fact that the inverters used in the routing resources are of larger sizes than those used in the logic resources; thus, they consume larger leakage.

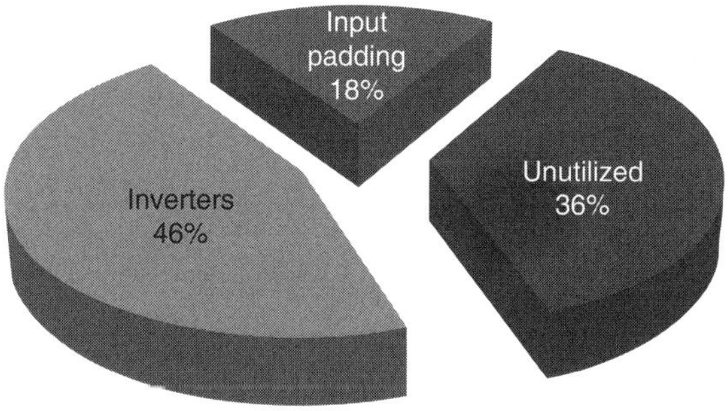

■ **FIGURE 6.11** Leakage savings breakdown in the routing resources.

6.3.1 **Pin Reordering and Performance**

The pin swapping algorithm results in changing the V_{TH} of the transistors in the pass-transistor multiplexers due to the DIBL and body effects previously explained. Its main aim is to have a net increase in V_{TH} to result in subthreshold leakage savings. As a result, the delay through these multiplexers ends up increasing. To investigate the impact of changing the input ordering on the design performance, several SPICE simulations were performed to calculate the average delay through the pass-transistor multiplexer for all the possible input combinations. A LUT is then structured for all the possible delays. To quantify the increase in delay due to the pin reordering algorithms, the delay across the critical path of the resulting designs are compared to those designed using the regular VPR and the performance penalty is plotted in Fig. 6.12. It can be seen that the performance penalty is always less 3% across all benchmarks tested.

In another experiment, the algorithm was applied in such a way to avoid changing the logic blocks and routing resources along the benchmark critical path. The critical path is identified using the state-dependent delay LUT explained above. Logic blocks and routing resources along the critical path are marked as "Do not touch" for the LPR and RPR algorithms not to change them. This version of the algorithm is called No-change for the Critical Path (NCP).

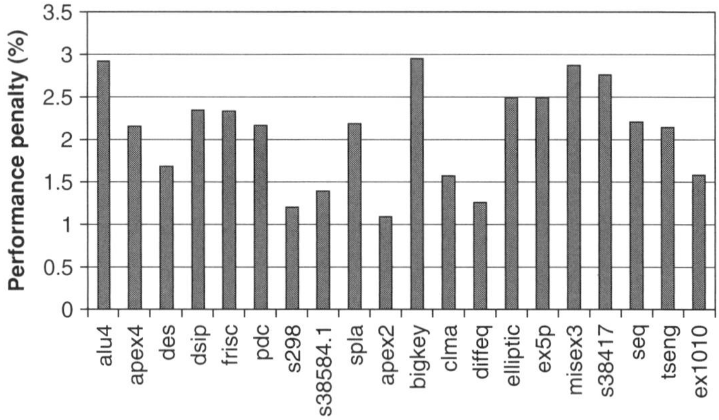

■ **FIGURE 6.12** Performance penalty due to the pin reordering algorithm.

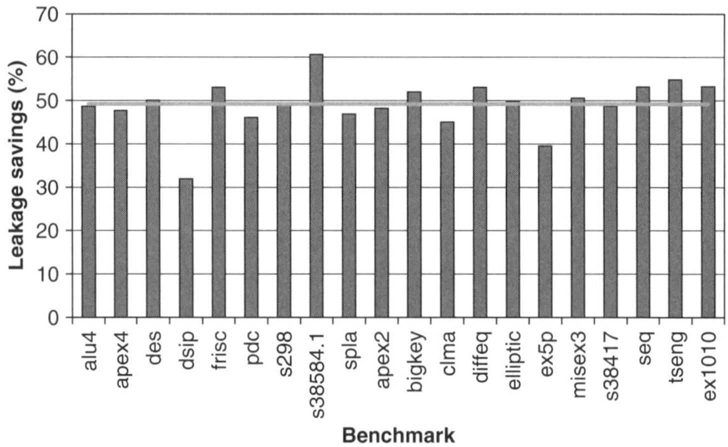

■ **FIGURE 6.13** Leakage savings to avoid affecting the performance.

The NCP version results in less leakage savings than those recorded in Table 6.2. The leakage savings resulting from the NCP versions are plotted in Fig. 6.13 for all the benchmarks tested. The horizontal line represents the average of the leakage reduction, which is around 49.14%, which is slightly less than the average savings in Table 6.2 (50.29%).

In another experiment, instead of leaving the logic and routing resources along the critical path unchanged, the pin swapping algorithm is applied to actually increase the multiplexer's V_{TH}. As a result, the leakage power along the critical path increases, while the delay along it is reduced. This provides a means of trading off some of the leakage savings to improve the circuit performance. This version of the algorithm is called Reduce the Critical Path (RCP). Figure 6.14 plots the percentage leakage savings as well

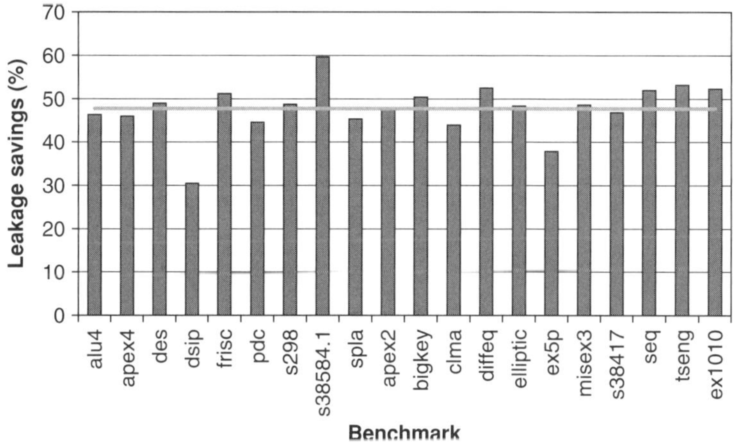

(a) Leakage savings for the RCP version.

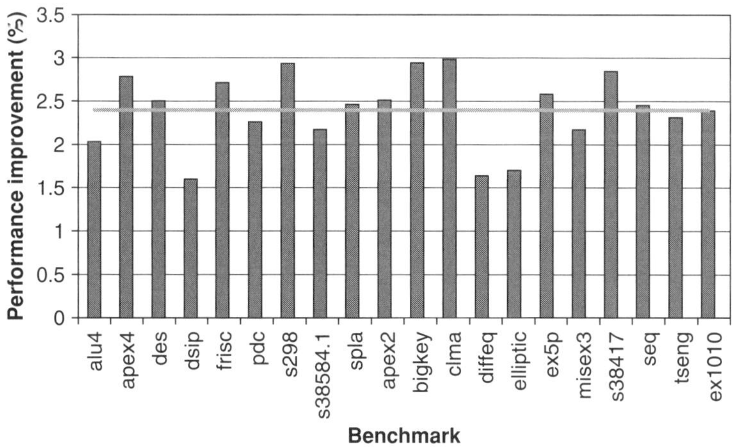

(b) Performance improvement for the RCP version.

■ **FIGURE 6.14** Trading leakage savings to reduce critical path delay in RCP.

as the improvement in the design performance for each benchmark achieved by the RCP version. The average performance improvement is 2.4%, whereas the average reduction in leakage savings is 47.7%.

6.3.2 Pin Reordering and Technology Scaling

Based on the ITRS report [41], both types of leakage currents are expected to increase significantly with the technology scaling as shown in Fig. 6.15. However, the increase in the gate leakage current is expected to be steeper, resulting in gate leakage current exceeding subthreshold leakage current in future CMOS technologies.

The minimum leakage state depends strongly on the relative magnitude of the subthreshold leakage and gate leakage currents. Hence, it is expected that the minimum leakage states will change for each technology. In the next set of experiments, the minimum leakage state is evaluated for several future technology nodes [143] using the BSIM4 leakage model. The results are presented in Fig. 6.16. It can be noticed that the state where the input and the output of the multiplexer are given by "0" is no longer the minimum

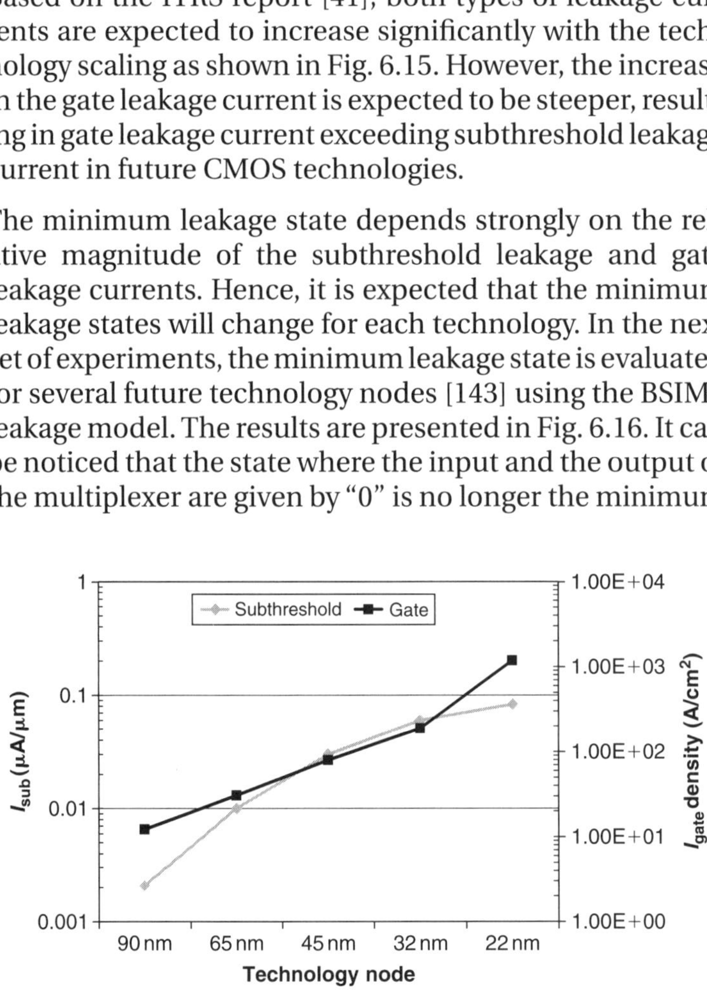

■ **FIGURE 6.15** Leakage current versus technology.

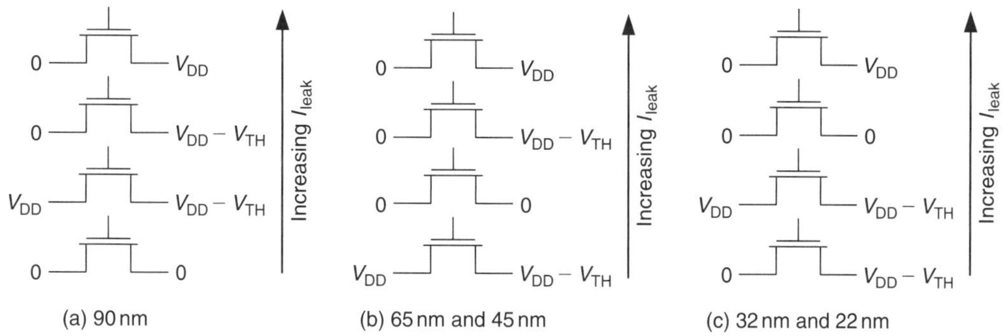

■ **FIGURE 6.16** Total leakage-dominant states in FPGA pass-transistor devices.

leakage state for technologies beyond the 90 nm. This is because this state has the maximum gate leakage current, so once the contribution of gate leakage current starts to dominate the total leakage current, the total leakage of that state will increase. The next observation is that as the technology is scaled down, the body effect starts to have a smaller effect on the device V_{TH}; hence, the total leakage of the state where the input and output are given by 1 starts to increase. Moreover, the gate leakage of that state is the second-highest gate leakage current achievable.

6.4 CONCLUSION

In this chapter, a leakage reduction algorithm is presented for FPGAs without any physical or performance penalties by using the input dependency of leakage power. Input reordering is used to place the logic and the routing resources in their lowest leakage state. The pin reordering methodology targets both the logic and the routing resources using the LPR and RPR algorithms, respectively. The newly developed algorithm achieves an average leakage savings of 50%. Another version of the algorithm is also developed, which results in a performance improvement of 2.4%, while achieving an average leakage reduction of 47.7%.

References

[1] Freeman RH. Configurable electrical circuit having configurable logic elements and configurable interconnects. U.S. Patent 4,870,302, Sept. 26, 1989.

[2] Zahiri B. Structured ASICs: opportunities and challenges. In: *Proceedings of the International Conference on Computer Design*. 2003:404–409.

[3] Taylor RR, Schmit H. Creating a power-aware structured ASIC. In: *Proceedings of the International Symposium on Low Power Electronics and Design*. 2004:74–77.

[4] Han KJ, Chan N, Kim S, et al. Flash-based field programmable gate array technology with deep trench isolation. In: *Proceedings of the IEEE Custom Integrated Circuits Conference*. 2007:89–91.

[5] Brown SD. An overview of technology, architecture and CAD tools for programmable logic devices. In: *Proceedings of the IEEE Custom Integrated Circuits Conference*. 1994:69–76.

[6] Greene J, Hamdy E, Beal S. Antifuse field programmable gate arrays. *Proc IEEE*. 1993;81(7):1042–1056.

[7] Ahmed E, Rose J. The effect of LUT and cluster size on deep-submicron FPGA performance and density. In: *Proceedings of the ACM International Symposium on Field Programmable Gate Arrays*. 2000:3–12.

[8] Rose J, Francis RJ, Lewis D, Chow P. Architecture of field-programmable gate arrays: the effect of logic block functionality on area efficiency. *IEEE J Solid-State Circuits*. 1990;25(5):1217–1225.

[9] Altera Corp. Stratix IV Device Handbook. [Online]. Available: http://www.altera.com/literature/hb/stratixiv/stratix4_handbook.pdf

[10] Xilinx Inc. Vertix-5 FPGA User Guide. [Online]. Available: http://www.xilinx.com/support/documentation/user_guides/ug190.pdf

[11] Actel Corp. ProASIC3 Handbook. [Online]. Available: http://www.actel.com/documents/PA3_HB.pdf

[12] Actel Corp. Axcelerator Family FPGAs Datasheet. [Online]. Available: http://www.actel.com/documents/AX_DS.pdf

[13] Rose J, Brown S. Flexibility of interconnection structures for field-programmable gate arrays. *IEEE J Solid-State Circuits*. 1991;26(3):277–282.

[14] Cong J, Ding Y. On area/depth trade-off in LUT-based FPGA technology mapping. In: *Proceedings of the IEEE/ACM Design Automation Conference*. 1993:213–218.

[15] Cong J, Peck J, Ding Y. Rasp: a general logic synthesis system for SRAM-based FPGAs. In: *Proceedings of the IEEE/ACM Design Automation Conference*. 1996:137–143.

[16] Cong J, Wu C, Ding Y. Cut ranking and pruning: enabling a general and efficient FPGA mapping solution. In: *Proceedings of the ACM International Symposium on Field Programmable Gate Arrays*. 1999:29–35.

[17] Ling A, Singh DP, Brown SD. FPGA technology mapping: a study of optimality. In: *Proceedings of the IEEE/ACM Design Automation Conference*. 2005:427–432.

[18] Marquardt A, Betz V, Rose J. Using cluster-based logic blocks and timing-driven packing to improve FPGA speed and density. In: *Proceedings of the ACM International Symposium on Field Programmable Gate Arrays*. 1999: 37–46.

[19] Cong J, Hargen L, Kahng AB. Random walks for circuit clustering. In: *Proceedings of the IEEE International Conference on Application Specific Integrated Circuits*. 1991: 14–21.

[20] Cong J, Lim SK. Edge separability based circuit clustering with application to circuit partitioning. In: *Proceedings of the IEEE/ACM Asia South Pacific Design Automation Conference*. 2000:429–434.

[21] Hagen LW, Kahng AB. Combining problem reduction and adaptive multi-start: a new technique for superior iterative partitioning. *IEEE Trans Comput Aided Des.* 1997;16(7):709–717.

[22] Huang DJ-H, Kahng AB. When clusters meet partitions: new density-based methods for circuit decomposition. In: *Proceedings of the European Design and Test Conference.* 1995:60–64.

[23] Dunlop AE, Kernighan BW. A procedure for placement of standard cell VLSI circuits. *IEEE Trans Comput Aided Des.* 1985;4(1):92–98.

[24] Huang DJ-H, Kahng AB. Partitioning-based standard-cell global placement with an exact objective. In: *Proceedings of the ACM International Symposium on Physical Design.* 1997:18–25.

[25] Kleinhans JM, Sigl G, Johannes FM, Antreich KJ. GORDIAN: VLSI placement by quadratic programming and slicing optimization. *IEEE Trans Comput Aided Des.* 1991;10(3):356–365.

[26] Srinivasan A, Chaudhary K, Kuh ES. Ritual: a performance driven placement algorithm for small cell ICs. In: *Proceedings of the International Conference on Computer Aided Design.* 1991:48–51.

[27] Marquardt A, Betz V, Rose J. Timing-driven placement for FPGAs. In: *Proceedings of the ACM International Symposium on Field Programmable Gate Arrays.* 2000: 203–213.

[28] Betz V, Rose J, Marquardt A. *Architecture and CAD for deep-submicron FPGAs.* Norwell, MA: Kluwer Academic Publishers; 1999.

[29] Sechen C, Sangiovanni-Vincentelli A. The TimberWolf placement and routing package. *IEEE J Solid-State Circuits.* 1985;20(4):510–522.

[30] Kirpatrick S, Gelatt CD, Vecchi MP. Optimization by simulated annealing. *Science.* Vol. 220, No. 4598, pp. 671–680, 1983.

[31] Hitchcock RB. Timing verification and the timing analysis program. In: *Proceedings of the IEEE/ACM Design Automation Conference*. 1982:594–604.

[32] Lewis D, Ahmed E, Baeckler G, et al. The Stratix II logic and routing architecture. In: *Proceedings of the ACM International Symposium on Field Programmable Gate Arrays*. 2005:14–20.

[33] Lewis D, Betz V, Jefferson D, et al. The Stratix™ routing and logic architecture. In: *Proceedings of the ACM International Symposium on Field Programmable Gate Arrays*. 2003:12–20.

[34] Lemieux GG, Brown SD. A detailed routing algorithm for allocating wire segments in field-programmable gate arrays. In: *Proceedings of the ACM Physical Design Workshop*. 1993:215–226.

[35] Chang Y-W, Wong DF, Wong CK. Universal switch modules for FPGA design. *ACM Trans Des Autom Electron Syst*. 1996;1(1):80–101.

[36] Wilton SJ. Architectures and algorithms for field-programmable gate arrays with embedded memory. Ph.D. dissertation, Univ. of Toronto, Toronto, 1997. [Online]. Available: http://www.eecg.toronto.edu/jayar/pubs/theses/Wilton/StevenWilton.pdf

[37] Masud MI, Wilton SJ. A new switch block for segmented FPGAs. In: *Proceedings of the International Workshop on Field Programmable Logic and Applications*. 1999:274–281.

[38] Ebeling C, McMurchie L, Hauck SA, Burns S. Placement and routing tools for the triptych FPGA. *IEEE Trans VLSI Syst*. 1995;3(4):473–482.

[39] Moore GE. Lithography and the future of Moore's law. In: *Proceedings of the Optical/Laser Microlithography VIII*. Vol. 2440, 1995:2–17.

[40] Moore GE. No exponential is forever: but "forever" can be delayed. In: *Digest of Technical Papers IEEE International Solid-State Circuits Conference*. 2003:2–17.

[41] The International Technology Roadmap for Semiconductors website. [Online]. Available: http://public.itrs.net

[42] Kuon I, Rose J. Measuring the gap between FPGAs and ASICs. *IEEE Trans Comput Aided Des.* 2007;26(2): 203–215.

[43] Roy K, Mukhopadhyay S, Mahmoodi-Meimand H. Leakage current mechanisms and leakage reduction techniques in deep-submicrometer CMOS circuits. *Proc IEEE.* 2003;91(2):305–327.

[44] Tuan T, Lai B. Leakage power analysis of a 90nm FPGA. In: *Proceedings of the IEEE Custom Integrated Circuits Conference.* 2003:57–60.

[45] Najm FN. A survey of power estimation techniques in VLSI circuits. *IEEE Trans VLSI Syst.* 1994;2(4):446–455.

[46] Vemuru SR, Scheinberg N. Short-circuit power dissipation estimation for CMOS logic gates. *IEEE Trans Circuits Syst I.* 1994;41(11):762–765.

[47] Wang Q, Vrudhula SB. On short circuit power estimation of CMOS inverters. In: *Proceedings of the International Conference on Computer Design.* 1998:70–75.

[48] Acar E, Arunachalam R, Nassif SR. Predicting short circuit power from timing models. In: *Proceedings of the IEEE/ACM Asia South Pacific Design Automation Conference.* 2003:277–282.

[49] Fatemi H, Nazarian S, Pedram M. A current-based method for short circuit power calculation under noisy input waveforms. In: *Proceedings of the IEEE/ACM Asia South Pacific Design Automation Conference.* 2007:774–779.

[50] Poon K, Wilton S, Yan A. A detailed power model for field-programmable gate arrays. *ACM Trans Des Autom Electron Syst.* 2005;10(2):279–302.

[51] Schneider P, Krishnamoorthy S. Effects of correlations on accuracy of power analysis - an experimental study. In: *Proceedings of the International Symposium on Low Power Electronics and Design.* 1996:113–116.

[52] Kang SM. Accurate simulation of power dissipation in VLSI circuits. *IEEE J Solid-State Circuits.* 1986;21:889–891.

[53] Yacoub GY, Ku WH. An accurate simulation technique for short-circuit power dissipation based on current component isolation. In: *Proceedings of the International Symposium on Circuits and Systems.* 1989:1157–1161.

[54] Krodel TH. PowerPlay-fast dynamic power estimation based on logic simulation. In: *Proceedings of the International Conference on Computer Design.* 1991:96–100.

[55] Rouatbi F, Haroun B, Al-Khalili AJ. Power estimation tool for sub-micron CMOS VLSI circuits. In: *Proceedings of the International Conference on Computer Aided Design.* 1992:204–209.

[56] Givargis TD, Vahid F, Henkel J. Trace-driven system-level power evaluation of system-on-a-chip peripheral cores. In: *Proceedings of the IEEE/ACM Asia South Pacific Design Automation Conference.* 2001:306–312.

[57] Murugavel AK, Ranganathan N. Petri net modeling of gate and interconnect delays for power estimation. In: *Proceedings of the IEEE/ACM Design Automation Conference.* 2002:455–460.

[58] Burch R, Najm FN, Yang P, Trick TN. A Monte Carlo approach for power estimation. *IEEE Trans VLSI Syst.* 1993;1(1):63–71.

[59] Xakellis MG, Najm FN. Statistical estimation of the switching activity in digital circuits. In: *Proceedings of the IEEE/ACM Design Automation Conference.* 1994:728–733.

[60] Hill AM, Kang S-MS. Determining accuracy bounds for simulation-based switching activity estimation. *IEEE Trans Comput Aided Des.* 1996;15(6):611–618.

[61] Marculescu R, Marculescu D, Pedram M. Sequence compaction for power estimation: theory and practice. *IEEE Trans Comput Aided Des.* 1999;18(7):973–993.

[62] Murugavel A, Ranganathan N, Chandramouli R, Chavali S. Least-square estimation of average power in digital CMOS circuits. *IEEE Trans VLSI Syst.* 2002;10(1):55–58.

[63] Liu X, Papaefthymiou MC. A Markov chain sequence generator for power macromodeling. *IEEE Trans Comput Aided Des.* 2004;23(7):1048–1062.

[64] Cirit MA. Estimating dynamic power consumption of CMOS circuits. In: *Proceedings of the International Conference on Computer Aided Design.* 1987:534–537.

[65] Najm FN. Transition density: a new measure of activity in digital circuits. *IEEE Trans Comput Aided Des.* 1993;12(2):310–323.

[66] Ghosh A, Devadas S, Keutzer K, White J. Estimation of average switching activity in combinational and sequential circuits. In: *Proceedings of the IEEE/ACM Design Automation Conference.* 1992:253–259.

[67] Marculescu R, Marculescu D, Pedram M. Efficient power estimation for highly correlated input streams. In: *Proceedings of the IEEE/ACM Design Automation Conference.* 1995:628–634.

[68] Marculescu R, Marculescu D, Pedram M. Probabilistic modeling of dependencies during switching activity analysis. *IEEE Trans Comput Aided Des.* 1998;17(2): 73–83.

[69] Tsui C-Y, Pedram M, Despain AM. Efficient estimation of dynamic power consumption under a real delay model. In: *Proceedings of the International Conference on Computer Aided Design.* 1993:224–228.

[70] Marculescu R, Marculescu D, Pedram M. Switching activity analysis considering spatiotemporal correlations. In: *Proceedings of the International Conference on Computer Aided Design.* 1994:294–299.

[71] Costa JC, Monteiro JC, Devadas S. Switching activity estimation using limited depth reconvergent path analysis. In: *Proceedings of the International Symposium on Low Power Electronics and Design.* 1997:184–189.

[72] Bhanja S, Ranganathan N. Switching activity estimation of VLSI circuits using bayesian networks. *IEEE Trans VLSI Syst.* 2003;11(4):558–567.

[73] Altera Corp. PowerPlay Early Power Estimator. [Online]. Available: http://www.altera.com/literature/ug/ug_stx3_epe.pdf

[74] Xilinx Inc. Xilinx Power Estimator User Guide. [Online]. Available: http://www.xilinx.com/products/design_resources/power_central/ug440.pdf

[75] Actel Corp. IGLOO Power Calculator. [Online]. Available: http://www.actel.com/documents/IGLOOpowercalculator.zip

[76] Altera Corp. PowerPlay Power Analysis. [Online]. Available: http://www.altera.com/literature/hb/qts/qts_qii53013.pdf

[77] Actel Corp. SmartPower v8.2 User's Guide. [Online]. Available: http://www.actel.com/documents/smartpower_ug.pdf

[78] Kusse E, Rabaey J. Low-energy embedded FPGA structures. In: *Proceedings of the International Symposium on Low Power Electronics and Design*. 1998:155–160.

[79] Weiß K, Oetker C, Katchan I, Steckstor T, Rosenstiel W. Power estimation approach for SRAM-based FPGAs. In: *Proceedings of the ACM International Symposium on Field Programmable Gate Arrays*. 2000:195–202.

[80] Shang L, Kaviani AS, Bathala K. Dynamic power consumption in Virtex^{TM}-II FPGA family. In: *Proceedings of the ACM International Symposium on Field Programmable Gate Arrays*. 2002:157–164.

[81] Degalahal V, Tuan T. Methodology for high level estimation of FPGA power consumption. In: *Proceedings of the IEEE/ACM Asia South Pacific Design Automation Conference*. 2005:657–660.

[82] Anderson JH, Najm FN. Power estimation techniques in FPGAs. *IEEE Trans VLSI Syst*. 2004;12(10):1015–1027.

[83] Poon K, Yan A, Wilton S. A flexible power model for FPGAs. In: *Proceedings of the ACM International Symposium on Field Programmable Gate Arrays*. 2002:312–321.

[84] Li F, Chen D, He L, Cong J. Architecture evaluation for power-efficient FPGAs. In: *Proceedings of the ACM International Symposium on Field Programmable Gate Arrays*. 2003:175–184.

[85] Li F, Lin Y, Lei H, Chen D, Cong J. Power modeling and characteristics of field programmable gate arrays. *IEEE Trans VLSI Syst*. 2005;24(11):1712–1724.

[86] Lin Y, Li F, He L. Power modeling and architecture evaluation for FPGAs with novel circuits for Vdd programmability. In: *Proceedings of the ACM International Symposium on Field Programmable Gate Arrays*. 2005:199–207.

[87] Kumar A, Anis M. An analytical state dependent leakage power model for FPGAs. In: *Proceedings of the Design, Automation, and Test in Europe*. 2006:612–617.

[88] Chou T-L, Roy K. Statistical estimation of sequential circuit activity. In: *Proceedings of the International Conference on Computer Aided Design*. 1995:34–37.

[89] Lui HY, Lee CH, Patel RH. Power estimation and thermal budgeting methodology for FPGAs. In: *Proceedings of the IEEE Custom Integrated Circuits Conference*. 2004:711–714.

[90] Hassan H, Anis M, Elmasry M. Total power modeling in FPGAs under spatial correlation. *IEEE Trans VLSI Syst*. 2009;17(4):578–582.

[91] Wunderlich H-J. PROTEST: a tool for probabilistic testability analysis. In: *Proceedings of the IEEE/ACM Design Automation Conference*. 1985:204–211.

[92] Seth SC, Pan L, Agrawal VD. PREDICT - probabilistic estimation of digital circuit testability. In: *Digest of Technical Papers IEEE International Symposium on Fault-Tolerant Comp*. 1985:220–225.

[93] Maamari F, Rajski J. A reconvergent fanout analysis for efficient exact fault simulation of combinational circuits. In: *Digest of Technical Papers IEEE International Symposium on Fault-Tolerant Comp*. 1988:122–127.

[94] Chakravarty S, Hunt HB. On computing signal probability and detection probability of stuck-at faults. *IEEE Trans Comput*. 1990;39(11):1369–1377.

[95] Tsui C-Y, Pedram M, Despain AM. Exact and approximate methods for calculating signal and transition probabilities in FSMs. In: *Proceedings of the IEEE/ACM Design Automation Conference*. 1994:18–23.

[96] Anderson J, Najm FN, Tuan T. Active leakage power optimization for FPGAs. In: *Proceedings of the ACM International Symposium on Field Programmable Gate Arrays*. 2004:33–41.

[97] George V, Zhang H, Rabaey J. The design of a low energy FPGA. In: *Proceedings of the International Symposium on Low Power Electronics and Design*. 1999:188–193.

[98] George V, Rabaey J. *Low-Energy FPGAs: Architecture and Design*. Boston, MA: Kluwer Academic Publishers, 2001.

[99] Li F, Lin Y, He L. Vdd programmability to reduce FPGA interconnect power. In: *Proceedings of the International Conference on Computer Aided Design*. 2004: 760–765.

[100] Li F, Lin Y, He L, Cong J. FPGA power reduction using configurable dual-Vdd. In: *Proceedings of the IEEE/ACM Design Automation Conference*. 2004:735–740.

[101] Li F, Lin Y, He L, Cong J. Low-power FPGA using pre-defined dual-Vdd/dual-Vt fabrics, In: *Proceedings of the ACM International Symposium on Field Programmable Gate Arrays*. 2004:42–50.

[102] Lin Y, Li F, He L. Routing track duplication with fine-grained power gating for FPGA interconnect power reduction. In: *Proceedings of the IEEE/ACM Asia South Pacific Design Automation Conference*. 2005:645–650.

[103] Hu Y, Lin Y, He L, Tuan T. Simultaneous time slack budgeting and retiming for dual-Vdd FPGA power reduction. In: *Proceedings of the IEEE/ACM Design Automation Conference*. 2006:478–483.

[104] Lin Y, Hu Y, He L, Raghunat V. An efficient chip-level time slack allocation algorithm for dual-Vdd FPGA power reduction. In: *Proceedings of the International Symposium on Low Power Electronics and Design*. 2006:168–173.

[105] Lin Y, He L. Statistical dual-Vdd assignment for FPGA interconnect power reduction. In: *Proceedings of the Design, Automation, and Test in Europe*. 2007:636–641.

[106] Lamoureux J, Lemieux GG, Wilton SJE. Glitchless: An Active Glitch Minimization Technique for FPGAs. In: *Proceedings of the ACM International Symposium on Field Programmable Gate Arrays*. 2007:156–165.

[107] Kim D, Choi K. Power-conscious high level synthesis using loop folding. In: *Proceedings of the IEEE/ACM Design Automation Conference*. 1997:441–445.

[108] Maymandi-Nejad M, Sachdev M. A digitally programmable delay element: design and analysis. *IEEE Trans VLSI Syst*. 2003;11(5):871–878.

[109] Maymandi-Nejad M, Sachdev M. A monotonic digitally controlled delay element. *IEEE J Solid-State Circuits*. 2005;40(11):2212–2219.

[110] Schell B, Tsividis Y. A low power tunable delay element suitable for asynchronous delays of burst information. *IEEE J Solid-State Circuits*. 2008;43(5):1227–1234.

[111] Lim H, Lee K, Cho Y, Chang N. Flip-flop insertion with shifted-phase clocks for FPGA power reduction. In: *Proceedings of the International Conference on Computer Aided Design*. 2005:335–342.

[112] Czajkowski TS, Brown SD. Using negative edge triggered FFs to reduce glitching power in FPGA circuits. In: *Proceedings of the IEEE/ACM Design Automation Conference*. 2007:324–329.

[113] Hsieh C-T, Cong J, Zhang Z, Chang S-C. Behavioral synthesis with activating unused flip-flops for reducing glitch power in FPGA. In: *Proceedings of the IEEE/ACM Asia South Pacific Design Automation Conference*. 2008: 10–15.

[114] Pedram M, Chang J. Module assignment for low power. In: *Proceedings of the Conference on European Design Automation*. 1996:376–381.

[115] Lyuh C-G, Kim T. High-level synthesis for low power based on network flow method. *IEEE Trans VLSI Syst*. 2003;11(3):364–375.

[116] Francis R, Rose J, Vranesic Z. Chortle-crf: fast technology mapping for lookup table-based FPGAs. In: *Proceedings of the IEEE/ACM Design Automation Conference*. 1991: 227–233.

[117] Cong J, Hwang Y-Y. Simultaneous depth and area minimization in LUT-based FPGA mapping. In: *Proceedings of the ACM International Symposium on Field Programmable Gate Arrays*. 1995:68–74.

[118] Kao C-C, Lai Y-T. An efficient algorithm for finding the minimal-area FPGA technology mapping. *ACM Trans Des Autom Electron Syst*. 2005;10(1):168–186.

[119] Cong J, Ding Y. FlowMap: an optimal technology mapping algorithm for delay optimization in lookup-table based FPGA designs. *IEEE Trans Comput Aided Des*. 1994;13(1):1–12.

[120] Cong J, Ding Y. On area/depth trade-off in LUT-based FPGA technology mapping. *IEEE Trans VLSI Syst*. 1994;2(2):137–148.

[121] Teslenko M, Dubrova E. Hermes: LUT FPGA technology mapping algorithm for area minimization with optimum depth. In: *Proceedings of the International Conference on Computer Aided Design*. 2004:748–751.

[122] Anderson J, Najm FN. Power-aware technology mapping for LUT-based FPGAs. In: *Proceedings of the IEEE International Conference on Field-Programmable Technology*. 2002:211–218.

[123] Lamoureux J, Wilton S. On the interaction between power-aware FPGA CAD algorithms. In: *Proceedings of the International Conference on Computer Aided Design*. 2003:701–708.

[124] Gupta S, Anderson J, Farragher L, Wang Q. CAD techniques for power optimization in Virex-5 FPGAs. In: *Proceedings of the IEEE Custom Integrated Circuits Conference.* 2007:85–88.

[125] Vorwerk K, Raman M, Dunoyer J, Chung Hsu Y, Kundu A, Kennings A. A technique for minimizing power during FPGA placement. In: *Proceedings of the International Conference on Field Programmable Logic and Applications.* 2008:233–238.

[126] Hassan H, El-Daher A, Anis M, Elmasry M. Activity packing in FPGAs for leakage power reduction. In: *Proceedings of the Design, Automation, and Test in Europe.* 2005:212–217.

[127] Hassan H, Anis M, Elmasry M. LAP: a logic activity packing methodology for leakage power-tolerant FPGAs. In: *Proceedings of the International Symposium on Low Power Electronics and Design.* 2005:257–262.

[128] Hassan H, Anis M, Elmasry M. A timing driven algorithm for leakage reduction in MTCMOS FPGAs. In: *Proceedings of the IEEE/ACM Asia South Pacific Design Automation Conference.* 2007:678–683.

[129] Gayasen A, Tsai Y, Vijaykrishnan N, Kandemir M, Irwin MJ, Tuan T. Reducing leakage energy in FPGAs using region-constrained placement. In: *Proceedings of the ACM International Symposium on Field Programmable Gate Arrays.* 2004:51–58.

[130] Kao J, Chandrakasan A. Dual-threshold voltage techniques for low-power digital circuits. *IEEE J Solid-State Circuits.* 2000;35(7):1009–1018.

[131] Mutoh S, Douseki T, Matsuya Y, Aoki T, Shigematsu S, Yamada J. 1-V power supply high-speed digital circuit technology with multithreshold-voltage CMOS. *IEEE J Solid-State Circuits.* 1995;30(8):847–854.

[132] Shigematsu S, Mutoh S, Tanabe YMY, Yamada J. A 1-V high-speed MTCMOS circuit scheme for power-down application circuits. *IEEE J Solid-State Circuits.* 1997;32(6): 861–869.

[133] Anis M, Areibi S, Elmasry M. Design and optimization of multithreshold CMOS (MTCMOS) circuits. *IEEE Trans Comput Aided Des*. 2003;22(10):1324–1342.

[134] Altera. Stratix Device Handbook, Volume 1. [Online]. Available: http://www.altera.com/literature/hb/stx/stratix_handbook.pdf

[135] Xilinx. Two flows for partial reconfiguration: module based or difference based. [Online]. Available: http://direct.xilinx.com/bvdocs/publications/xapp290.pdf

[136] Hu Z, Buyuktosunoglu A, Srinivasan V, Zyuban V, Jacobson H, Bose P. Microarchitectural techniques for power gating of execution units. In: *Proceedings of the International Symposium on Low Power Electronics and Design*. 2004:32–37.

[137] Rahman A, Das S, Tuan T, Trimberger S. Determination of power gating granularity for FPGA fabric. In: *Proceedings of the IEEE Custom Integrated Circuits Conference*. 2006: 9–12.

[138] Kosonocky SV, Immediato M, Cottrell P, Hook T, Mann R, Brown J. Enchanced multi-threshold (MTCMOS) circuits using variable well bias. In: *Proceedings of the International Symposium on Low Power Electronics and Design*. 2001:165–169.

[139] Kim H-O, Shin Y, Kim H, Eo I. Physical design methodology of power gating circuits for standard-cell-based design. In: *Proceedings of the IEEE/ACM Design Automation Conference*. 2006:109–112.

[140] Calhoun B, Honoré F, Chandrakasan A. A leakage reduction methodology for distributed MTCMOS. *IEEE J Solid-State Circuits*. 2004;39(5):818–826.

[141] Tuan T, Kao S, Rahman A, Das S, Trimberger S. A 90nm low-power FPGA for battery-powered applications. In: *Proceedings of the ACM International Symposium on Field Programmable Gate Arrays*. 2006:3–11.

[142] James Kao DA, Chandrakasan A. Transistor sizing issues and tool for multi-threshold CMOS technology. In: *Proceedings of the IEEE/ACM Design Automation Conference.* 1997:409–414.

[143] The Berkeley Predictive Technology Model website. 2008 [Online]. Available: http://www-device.eecs.berkeley.edu/~ptm/

[144] Guindi RS, Najm FN. Design techniques for gate-leakage reduction in CMOS circuits. In: *Proceedings of the IEEE International Symposium on Quality of Electronic Design.* 2003:61–65.

[145] Hassan H, Anis M, Elmasry M. Input vector reordering for leakage power reduction in FPGAs. *IEEE Trans Comput Aided Design.* 2008;27(9):1555–1564.

[146] Sentovich EM, Singh KJ, Lavagno L, et al. SIS: A system for sequential circuit synthesis. Department of Electrical Engineering and Computer Science, University of California, Berkeley, CA, Technical Report UCB/ERL M92/41, May 1992.

Index